TELEVISION PRODUCING & DIRECTING

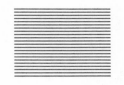

Howard J.
Blumenthal

TELEVISION PRODUCING
& DIRECTING

Barnes & Noble Books
A Division of Harper & Row, Publishers
New York, Cambridge, Philadelphia, San Francisco, Washington
London, Mexico City, São Paulo, Singapore, Sydney

FIRST EDITION

Designer: Laura Hough

Library of Congress Cataloging-in-Publication Data

Blumenthal, Howard J.
　Television producing and directing.
　(Everyday handbook; EH/700)
　Bibliography: p.
　Includes index.
　1. Television—Production and direction.　I. Title.
PN1992.75.B58　1987　　　791.45′023　　　86-45080
ISBN 0-06-463700-X (pbk.)

87 88 89 90 91 RRD 10 9 8 7 6 5 4 3 2 1

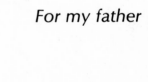

For my father

CONTENTS

PREFACE

This is a book about television production. It is written for anyone who wants to know what producing and directing are like in the real world of television. It will not teach you how to operate a television camera, nor how to light a set yourself (these are crafts assigned to skilled technicians and are not the direct responsibility of the producer or the director). Instead, it will concentrate on the process involved in creating a production, and on the qualities of leadership that are essential to the making of any television project, regardless of its size, regardless of its intended audience, regardless of its budget. In a word, this book is about being a professional.

The book is organized very much like a production. It begins with a consideration of the markets (chapter 1), and of the jobs available in those markets (chapter 2). Since every production begins with an idea, the process of transforming that idea into a salable project is described for both fiction (e.g., prime time, music videos) and nonfiction (e.g., news and sports, educational and corporate presentations, home video programs), from proposals and treatments through budgets (chapter 3). Once the idea has been approved, the planning begins, staff is hired, and a schedule is written (chapter 4).

The middle of the book is devoted to shooting, first in the field (chapter 5), then in the studio (chapter 6). More advanced studio production, special effects, and large-scale remote production are also described in detail (chapter 7). Finally, it explains how the production is edited and how a sound track is added (chapter 8).

The producer, director, performers, and crew are not the only ones responsible for a successful production. Every production is a group effort, involving set designers, graphic artists, wardrobe and makeup artists, composers and musicians (chapter 9). The business side of production, from employment contracts to union affiliations, is one of the book's most important components (chapter 10). This discussion of the business side prompts a fresh look at the real world, the very real problems in finding a first job and in making changes later in a career (chapter 11).

There is a glossary, which you may decide to update as the need arises, and some suggestions for further reading. The reading-suggestions section includes a few solid reference works and textbooks, but mainly it's a list of relevant trade magazines that will keep you posted as to the latest news and trends in all parts of the television production industry.

You won't become a successful producer or director by reading this book. You must learn the techniques from actual experience. But you'll be well on your way. You'll have a clear understanding of how producers and directors make television productions, and an equally clear understanding of all that must be done before somebody decides to trust you and says, "I'd like you to produce it for me" or "You've been watching me long enough. Get into the chair and direct it yourself!"

I hope this book will help.

New York City, 1986

ACKNOWLEDGMENTS

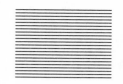

This book is an accumulation of knowledge and good advice, much of it provided by professionals with whom I have had the pleasure to work. The contributions of Diane Vilagi, Pat Cocco, Ariel Schwartz, Bob Small, Hal Rein, David McGoldrick, Robin McClellan, Marty Bell, Barry Rebo, Dean Winkler and the staff of VCA Teletronics in New York are evident throughout this book, in some very specific and in many very general ways. To Al Markim, Harlan Kleiman, Dr. Vivian Horner, Michael Marcovsky, and all of the other people who gave me the opportunities to produce, and to build a career in this business, my most sincere and grateful thanks. For specific contributions, such as interviews and photo materials, my thanks to Tory Baker, Phyllis McGrady, Jaclyn Demave, George Barimo, Merrill Grant, Jane Lipstone, Don Carney, John Murray, Norm Carrier, and Amy Celniker of Lippin & Grant, as well as Julian Goldberg and Ann Marie Hancock. To good friends, like Robert Morton, Sue Steinberg, Marcy Dubrow, Nick Mitsos, Mike Fields, and so many others from the days of QUBE, I hope this book shows how much you have each taught me in your own special ways.

For Dorothy Curley, who not only read the entire manuscript in its several versions, but also contributed much of her own professional commentary and real-world experience, a very special thank-you.

For my wife, Shari, who helped me, provided support and encouragement throughout the several years that this book required, and is always there when I need her, my usual thanks.

And, finally, three professional thank-yous. The first, to my agent, Jeanne Glass, who understood the need for a book like this one. The second, to my editor, Jeanne Flagg, who managed to coax the best possible work out of the author. The third, to my father, to whom this book is dedicated, for his clear vision, his fantastic store of experience and knowledge, his many years of teaching me, and for the many hours spent on every page of the manuscript.

TELEVISION PRODUCING & DIRECTING

ONE

The Production Business

A few years ago, "working in television" meant, simply, employment at a local television station or at one of the networks. Television is changing rapidly, and with these changes has come a great variety of programming and production opportunities.

- Most large corporations, and many smaller ones, have their own television facilities, including studios and editing equipment.
- With the popularity of home videocassette recorders, programs are being produced exclusively for sale to consumers.
- Almost every important new record release is now accompanied by a music video, because music videos help to sell records.
- The syndication marketplace is booming, and as the cable industry stabilizes and the networks and local stations react to changes in the traditional ways of doing business, the need increases for producers, directors, writers, and other television production personnel.

Until recently, most of the "good" television jobs were to be found only in Los Angeles, New York City, Chicago, Toronto, and a few other large cities. It was common to find a producer who had lived in five cities in five years, in progressively larger markets. Now, there is work in every major city and in many smaller ones that are home to large corporations. It is possible to work your way up in one or two cities, or to start in a relatively modest position in a small city, in a corporate facility, with hopes of growing with the department. This is a very new situation for television production.

Also in the past, producers or directors were expected to select a single specialty and to stick with that specialty for a long while. Now, versatility is the key to long-term success. The ability to handle a corporate presentation or a major news program, a music video or a variety show, a field report or a studio piece, is becoming increasingly common. Which is not to say that everybody should be able to do everything: it is wise to specialize in information, or in entertainment, or in instruction, or in music. But within those broad areas, a producer or director can now do it all.

The television production marketplace has become quite large and somewhat complicated. This chapter is concerned with an analysis of the production marketplace.

There are ten basic subdivisions in the television production business:

1. Commercial broadcast networks
2. Broadcast syndication
3. Local stations
4. Cable and satellite networks
5. Cable systems
6. Educational television/video
7. Corporate video
8. Home video
9. Commercials
10. Music video

As television evolves, the distinctions among these will dissolve. Some educational outlets are currently using music videos to teach spelling; some programs originally produced for internal corporate use will be released as home videocassettes. It is best to think of these subdivisions as temporary and flexible.

Prime Time Show in Production
Single mothers Susan Saint James (Kate) and Jane Curtin (Allie) in a scene from their hit CBS situation comedy, ''Kate & Allie.'' It is one of a very small number of successful network prime time comedies produced in New York City. *(Photo courtesy Alan Landsburg Productions, Inc., Copyright 1985)*

COMMERCIAL BROADCAST NETWORKS (ABC, NBC, AND CBS)

1. Prime Time

With very few exceptions, the commercial broadcast networks produce programs in the Los Angeles area and in New York City. Nearly all entertainment programs are produced in LA, where the movie studios (MCA/Universal, Paramount, Columbia Pictures, Warner Brothers, Embassy, Walt Disney Productions, MGM/UA, and Twentieth Century-Fox) and the large production companies (Lorimar-Telepictures, MTM, Aaron Spelling, and others) have access to an established talent pool and the necessary stage and technical facilities. These companies develop most of the programs that are produced on network television because they have "track records" (industry lingo for "a string of successful projects"), as well as the money and the management skills needed to maintain a pool of top producers and writers.

Most prime time network projects begin with a writer, or, more often, a writer/producer. To maintain a steady flow of income, the writer/producer signs an agreement with a studio to develop one or more projects. The studio usually pays the writer/producer a fee during the development period in exchange for the right to co-own the properties and to most of the profits generated by some or all of the properties developed. Each of the studios maintains a staff of development people who are in constant touch with the network program buyers. The writer/producer works closely with the network buyers as well.

The development process, which is described in detail in chapter 3, is not only a matter of creating an idea. Most of the time, the networks don't buy ideas. They buy "packages."

Television program suppliers, called packagers, begin by developing a basic script or story line. Then the process of packaging begins. The most important part of the package is the "talent"—one or two stars who have appeared on network television before. The names of the director, the producer, and the writers (there may be more than one), even the director of photography and the art director, all strengthen the package.

A team of network programming executives considers the packages submitted over the course of the season, and decides to test a few of the concepts by committing money, first, for further script and package development, and, second, to produce one or more pilot episodes. The pilot is then screened for test audiences. Questionnaires reveal likes and dislikes, while sophisticated equipment measures reactions to specific scenes and characters.

A programming decision is not made by research alone, however. Strategies are developed based on available time slots (certain programs will only "work" on Sunday nights at 8:00 P.M., a traditional family viewing period), upon lead-ins (the program immediately preceding the open slot), and upon the existing competition (if ABC is programming a show that does well with teens, NBC would logically counterprogram with a show that would appeal to adults). These factors, plus the track record of the producers and the stars, and the ability of a program executive to "sell" his or her peers on a new concept, determine what is seen on prime time network television. Most projects don't make it past the piloting stage. And most new programs, even if they do make it to the fall (or any of the supplemental) schedules, remain on the air for only a few months before they are replaced by other bright newcomers.

2. Daytime and Other Times

Although most people think of network television in terms of prime time, the networks actually offer programs to local affiliates throughout most of the day and night. A typical schedule for the "network feed" (i.e., the schedule fed to local affiliates via telephone lines, microwaves, or satellite systems) looks something like this:

6:30 A.M.–7:00 A.M.	Early news
7:00 A.M.–9:00 A.M.	Morning programming (e.g., "Good Morning America")
9:00 A.M.–10:30 A.M.	(No network feed; local programming)
10:30 A.M.–4:00 P.M.	Daytime programming (e.g., soap operas and game shows)
4:00 P.M.–7:00 P.M.	(No network feed; local programming)
7:00 P.M.–7:30 P.M.	Network news*
8:00 P.M.–11:00 P.M.	Prime time
11:00 P.M.–11:30 P.M.	(No network feed; local news)
11:30 P.M.–1:00 A.M.	Late night programming
1:00 A.M.–Sign-off	(No network feed)

This schedule will vary slightly. The networks have been experimenting with late-night talk shows, all-night newscasts, and other innovations. They have gradually been increasing the number of daytime hours. There is talk of an hour-long network news block from 6:00 P.M. to 7:00 P.M. or 7:00 P.M. to 8:00 P.M. The basic chunks are likely to remain.

Morning programming in the style of "Today" and "Good Morning America" will continue to be produced by the networks. (Since these shows are image-builders for the networks, they are likely to remain under network control.) Like the network news programs, these are produced live in New York,

and, because of the time zone difference, either delayed-broadcast (with updates for news and weather), or, in some cases, produced for a second live broadcast for the West Coast. Each of these morning programs employs an enormous staff of producers and production personnel (if you'd like to see the complete list, watch on a Friday).

Daytime programming is almost always a mix of soap operas and game shows, with the occasional experiment in talk and variety shows, and some reruns to complete the schedule. The soaps are produced on both coasts, sometimes by advertising agencies, sometimes by program packagers. Game shows are produced, almost exclusively, in Los Angeles, usually by experienced game show hands like Mark Goodson ("Family Feud," "The Price Is Right"), Merv Griffin ("Wheel of Fortune," "Jeopardy"), and Bob Stewart ("$25,000 Pyramid"). As in prime time, game show packagers develop a basic concept, create a package, and present the idea to network programming executives. Additional money for development, a commitment to make a pilot, and the inevitable "testing" are the next steps. Most game shows, like their prime time equivalents, have remarkably short runs. Since the networks are constantly replacing failed game shows with new ones, and game shows cost less to develop than prime time shows, many small production companies in LA (and NY) develop game show ideas. Most are rejected by the networks in the early stages, but a few result in development deals and pilots. A very small percentage get on the air, usually for a short period of time.

When a soap, or "daytime drama," is successful, its life-span may be measured in decades. While a few new soaps have made it in recent years, programs like "General Hospital" and "The Guiding Light" consistently dominate daytime schedules. Developing a soap is an extremely complex process, even

---

* Network news is usually fed at 6:30 P.M. and again at 7:00 P.M. Local stations opt for the live (6:30 P.M.) or tape (7:00 P.M.) feed, knowing that the later feed will be updated by the network for the late-breaking stories and updates.

"Good Morning America" in Production
Host David Hartman takes a five-second cue from the stage manager on "Good Morning America." The atmosphere appears to be quite relaxed, despite the pressure to create a live two-hour program every weekday morning, and to win top ratings as well. *(Photo courtesy American Broadcasting Company © 1985)*

more complicated than prime time development, because the daily schedule demands a few dozen characters and as many situations. Only a few producers have mastered it. Opportunities tend to be limited here because production companies usually hire people experienced in soaps, and because there are so few new soaps every year.

The evening news program is produced by the network, again by a large number of staff people. Most either worked their way up within the network news organization or came from local news operations around the country. You'll find more about news production later in the book.

As for late night programming, it's a mix. As of this writing, ABC runs "Nightline," which is produced in Washington, D.C., by ABC News, CBS runs a movie, reruns, or an original action-adventure, and NBC runs "The Tonight Show Starring Johnny Carson" (produced by Carson Productions), followed by "Late Night with David Letterman" (co-produced by Carson Productions and Rollins-Joffe, a company best known for producing Woody Allen's movies). Through the past ten years, as late night programming has been developing, some programs have been produced by network staffs while others have been produced by outside production companies. Most of the programs come out of LA, but New York has always seen its share of the action. "Late Night with David Letterman" and its predecessor, "Tomorrow" with Tom Snyder, were produced in New York. "Saturday Night Live" has long been associated with New York as well.

Live sports programming occupies many weekend hours on all three networks. Each of the networks employs a sports department, with staff producers, directors, even graphic designers. Coverage is generally preproduced in New York, which is where the production offices are located, and where programs like "NBC Sports Saturday" and "ABC's Wide World of Sports" are produced.

BROADCAST NETWORKS (PBS)

The structure of the Public Broadcasting Service, or PBS, differs from that of the commercial networks in several ways. Although PBS feeds a nightly lineup of programs, most are produced not by the network, but by the largest local stations in the system. Stations like WGBH (Boston), WNET (New York), WQED (Pittsburgh), and WTTW (Chicago) are among the most prolific suppliers to the PBS schedule. WGBH, for example, has provided PBS stations with "Masterpiece Theatre," "Mystery," "This Old House," "Nova," and "Evening at Pops." WQED has been responsible for "National Geographic Specials" and "Mister Rogers' Neighborhood." WTTW has produced "Soundstage" and "Sneak Previews." Some stations, like WGBH, Boston, produce many hours of programming, season after season. Others may turn out only a few hours' worth every few seasons, because of limited funding, limited facilities, management's tendency to channel energies into local programming, or an erratic flow of program concepts that are suitable for the network sched-

ule. Each winter, PBS affiliates gather at a "Program Fair" to review the next season's offerings. PBS also maintains a national programming office in Washington, D.C., which acquires completed programs but does not usually initiate production, for this is seen as the responsibility of the local stations. Several regional networks, like the Eastern Educational Network (EEN), arrange smaller chains of PBS stations for specific projects. This is possible with public television because of a relatively flexible relationship between local stations and the PBS network.

Because of PBS, it is possible for a producer or director to create high-quality network programming while living in Pittsburgh, San Francisco, Chicago, or Boston. Most PBS stations maintain a basic programming and production staff whose primary responsibility is the local schedule, usually talk shows and some documentaries. In most places, this staff is also responsible for the development of corporate contributions and for any PBS network properties that may logically originate from their facilities. In the larger PBS originating stations, a full staff is employed for the development, financing, and production of network programs.

Most PBS programs are financed by grants from foundations (like the Ford Foundation), corporations (like Mobil), and local station budgets. The Corporation for Public Broadcasting also provides partial financing on many projects. Corporations based in large cities usually support the local PBS station, both in their operations budgets and on specific program projects.

BROADCAST SYNDICATION

Whether a local television station is a network affiliate or an independent, the name of the game is ratings. When ratings are high, more people are watching, so advertisers pay top dollar for commercials. The best way to insure high ratings is to schedule programs with a guaranteed audience. Because network programs are inevitably the highest-rated shows, network affiliates almost always have higher ratings than their nonnetwork competitors (there are a few notable exceptions, like major sporting events for winning teams). In those time slots when the network does not schedule programs, high ratings are most easily obtained by paying a fee to a syndication company, like Viacom, for the exclusive right to play, for example, "Dallas" or "M*A*S*H*" in a specific market. (A "market," also known as an ADI for "area of dominant influence," is the area where the station's signal is clear and easily received with only a small home antenna.)

Most of the programs that are syndicated to local stations are off-network reruns (i.e., programs that have been seen previously as prime time entries on a commercial broadcast network). Each winter at the NATPE (National Association of Television Program Executives) convention, the syndicators offer stations a wide array of off-network series, new programs, movies, and repackaged versions of old programs that will be available for the following fall season. Only one station in each market can buy a show; when one station in a market has agreed to buy, the market is said to be "cleared." If the syndicator can clear most of the top twenty markets (number one is New York; number twenty is usually a city like Miami, San Diego, or Cleveland, depending on the year and who's doing the counting), as well as a total of 75 to 100 of the top 200 to 250 stations, the show will be syndicated, and, if the show is an original, production will begin. If the program is not "cleared" on a sufficient number of stations, it does not go on the air. Successful programs produced for syndication appear throughout independent stations' schedules and in the pre-daytime and "prime

access" (pre-prime time) slots on network stations. "The Newlywed Game," "Entertainment Tonight," "Fame," "Lifestyles of the Rich and Famous," and the evening versions of network game shows like "Family Feud" and "Wheel of Fortune" are all examples of programs produced for the syndication marketplace.

Viacom, Lorimar-Telepictures, and Paramount are among the most active television syndicators. Most of the movie studios have syndication arms. There are many smaller syndicators as well. Note that most syndicated programs are produced in LA and NY, but some are produced elsewhere. "The Phil Donahue Show" was produced for many years from Chicago, and, more recently, "Sally Jesse Rafael" is produced in St. Louis.

LOCAL STATIONS

There are three kinds of local television stations: network affiliates, public television stations, and independents ("indies").

1. Network Affiliates

As a rule, network affiliates carry the entire network schedule and fill the additional hours with local news, a few locally produced talk shows (to fulfill a minimum number of hours of "local public service" required in their operator's license), and, in some cases, local sports presentations. Network affiliates are usually the largest, most successful stations in a market. Also, network affiliates usually carry the top syndicated shows because they have the money to pay for them. With the increasingly powerful station group owners (such as Post-Newsweek, Gannett, Storer, Outlet, and Westinghouse) making buy decisions for as many as seven stations, this situation is chang-

Local News in Production
The evening news at WOKR, Channel 13 in Rochester, New York. The program is produced with a relatively small staff, shot with three cameras (only the two seen here are manned; the third camera is locked in position). Note the shot sheet (the list of camera shots) on the back of camera 2, just below the viewfinder.

ing rapidly. Each network owns up to seven stations in major markets. NBC, for example, owns WNBC in New York, WMAQ in Chicago, and WRC in Washington, D.C. CBS owns WCBS in New York and WCAU in Philadelphia. ABC, or, more accurately, ABC/Capital Cities, owns WABC in New York, KABC in Los Angeles, and WLS in Chicago.

Local news programs have turned out to be quite profitable for most local stations, so it is here that local stations spend most of their production dollars. The largest cities have news operations with staffs numbering over one hundred people. Stations in the smallest cities employ a dozen or more people in the production of local news. Very often, the local news staff is responsible for a Sunday-morning public affairs/talk show as well.

Through the late 1970s and early 1980s, many local stations were quite successful with a nightly "magazine" show (e.g., "PM Magazine"). Although audience interest in these programs is on the wane, stations have already made the investment in portable video equipment and editing rooms for these magazine shows. Many local news operations have already absorbed the equipment.

2. Public Television Stations

Public television stations usually carry most of the PBS feed, and supplement with programs that are not carried on the network. Most public television stations also produce some original programming: the larger ones produce mainly prime time programs for the network, while the smaller ones produce the educational programming seen during the daytime hours.

Public television stations outside the largest markets are frequently owned and/or operated by colleges and universities.

3. Independent Stations

Most of the top twenty-five markets have at least five stations: three network affiliates, one PBS station, and one independent station (or "indie"). Indies are frequently owned by "station groups,"* like Tribune or Gannett, which buy programming on behalf of all of their stations.

Independent stations have typically filled their schedules with movies, older syndicated programs (which are less expensive than the new ones), and sporting events. With no commitment to a network for the evening hours, indies have the flexibility to schedule baseball games, for example, during prime time. These

---

* FCC regulations presently limit group ownership to seven television stations, seven AM and seven FM stations. As of this writing, this ruling is under consideration by the FCC, and may be changed in the near future.

games are usually produced by the stations as a result of long-term contracts with the teams. New York's WPIX, for example, has been carrying the New York Yankees games for many years. WOR, another New York independent, carries the Mets. New York's third independent station, WNYW is owned by Fox, so it carries programs produced in Los Angeles for the Fox network.

Independent stations may also form a temporary network for the nationwide airing of a special show or series (frequently a miniseries like "A Woman Named Golda" or "Blood Feud"). The success of the Operation Prime Time (OPT) and other ad-hoc networks has come about as a result of interest on the part of independent stations to compete with the network affiliates during the prime time hours. OPT programs usually carry big names and sensational story lines to attract the largest possible audience.

CABLE NETWORKS

With the easing of government regulations, the cable television business grew quickly in the 1970s. Large corporations, realizing that the ownership of cable systems would someday prove to be as profitable as the ownership of local television stations, started buying existing cable systems and bidding to build systems in cities where none existed. A rapid period of growth caused skyrocketing subscriber counts, and optimistic business projections encouraged the development of new cable networks. The most successful of these ventures, MTV, and Ted Turner's CNN, have redefined the concept of a television network.

Cable News Network, headquartered in Atlanta, Georgia, runs a twenty-four-hour-a-day, seven-days-a-week operation, with a large staff of producers, directors, writers, graphic artists, and, of course, reporters. CNN maintains bureaus in New York, Washington, and a few other cities, many quite large.

Home Box Office, headquartered in New York, depends upon movies for most of its schedule. HBO is extremely involved in the entertainment business, mainly in the financing of new films. It looks to the motion picture and television community in Los Angeles for most of its nonmovie production. In these instances, HBO's operations mirror the activities of the broadcast networks, with the studios and large production companies most actively involved in new project development. HBO

CNN in Production
CNN's headquarters in Atlanta, Georgia, employs several hundred people as newscasters, commentators, newswriters, news producers, graphic artists, directors, and technical personnel. The channel runs 24 hours a day, 365 days a year, and operates several smaller versions of this "open newsroom" in major cities like New York and London. *(Photo courtesy Cable News Network)*

also produces "continuity" material and some original information programming in New York (sports talk shows, "HBO Mailbox," and other minor efforts).

Cinemax, which is owned by HBO, is almost exclusively a movie service. It depends upon the established production companies for a limited amount of original programming, which is developed to fit the needs of Cinemax, and then financed, at least in part, by the service.

Showtime, HBO's most direct competition, also depends on movies to fill its schedule. Most of the original programming on Showtime comes from the movie studios and from large production companies who do business with the commercial networks as well. Showtime also produces a small amount of between-the-movies ("interstitial") programming.

American Movie Classics, owned by Cablevision (a large operator of cable systems) specializes in vintage films. The channel produces wraparounds and promos, but most of the airtime is filled, as you'd expect, with movies. It's located on Long Island.

Bravo, also owned by Cablevision, features acquired arts programming and foreign films. Besides promos and continuity material, there is almost no original production on the channel or programs.

Nickelodeon is owned by MTV Networks, which is in turn owned by Viacom. It is a children's channel with many hours of original programming in its weekly schedule. Some of the programs, like "Pinwheel," are produced by the network with a staff of free-lancers (program segments are videotaped, then replayed numerous times, so the staff is reunited only when the network needs new segments, such as "You Can't Do That on Television" and "Turkey Television"). Independent production companies of all sizes work with the Nickelodeon programmers to create original programming for children. The "Nick at Nite" schedule is reruns of classic programs, like

"My Three Sons," for the entire family. Financing comes from the network and from commerical sponsors.

MTV/Music Television has a program schedule that resembles radio. The network operates a "home base" studio in New York, where VJs prerecord their commentary several days in advance of air. This commentary is intercut with music videos, which are supplied by the record companies. MTV also produces, finances, and acquires special programming, usually concerts.

VH-1, also owned by MTV Networks, follows a similar format for an older audience. The channel depends almost entirely upon VJ's and music videos, with little original production.

Lifetime, based in New York and owned by ABC/Hearst and Viacom, runs a full day's schedule with many hours of original programming, and, to fill the schedule, some movies. Original productions are produced by outside production companies and financed by Lifetime and its sponsors. The orientation is women and contemporary family life.

The USA Network runs several different types of programming throughout the day and night. Most evenings are devoted to sports, and most of these presentations are produced by staffers at the network or by a cadre of free-lancers. The daytime schedule is principally acquisitions of existing shows, with some new productions as well. These productions, financed by sponsors, tend to run in two directions: women's services and rock music. The USA Network is headquartered in Manhattan.

ESPN was originally conceived as a twenty-four-hour sports channel, with live coverage (and replays) of sporting events filling the day's schedule. In addition, the network has been successful with business news programming. It's located in Connecticut.

The Nashville Network fills each day with a considerable amount of original programming, including talk shows ("Nashville Now"), game shows ("Fandango"), and specials. Just about

all of this country-oriented programming is produced in Nashville.

CBN, out of Virginia Beach, Virginia, offers network service twenty-four hours each day. Although much of the program schedule is filled with movies and reruns, the network does produce several talk shows and other programming on a regular basis and maintains a staff of producers, writers, directors, and performers. There is a Christian orientation to some of the programming.

The Disney Channel works with outside production companies for several different series and specials, using these programs alongside the many hours of programming available from the Disney library and programs produced by the studio.

The Playboy Channel has changed its program orientation several times, and is now offering a combination of entertainment specials and adult movies. It does produce some original programs, like " . . . The News According to Playboy" and "Women on Sex."

The Weather Channel, also headquartered in Atlanta, offers twenty-four-hour information and forecasts. The production is very simple, but the channel does maintain a small staff of producers, directors, writers, researchers, and assistants. It is owned by Landmark Communications.

There are several shop-at-home channels as well. Although the emphasis is on selling products—and not on production or programming excellence—organizations like "Home Shopping Network" do employ producers and directors, and may present an opportunity, particularly at the early stages of a career.

Since the cable business is changing, it is best to spend some time watching these and other channels in order to gain a clear understanding of their present schedules and production activities, and to review cable guides for more information.

CABLE SYSTEMS

There are over four thousand local cable systems in the United States. Most are retransmission operations, whose small engineering staffs simply monitor incoming feeds from satellites (HBO, Showtime, CNN, and most of the other cable networks are distributed via satellite, so most cable systems have access to a large dish-type satellite "downlink"—a satellite antenna), and from microwaves (feeds of local stations from nearby cities are transmitted by microwave). These feeds are retransmitted by cable to households who pay a monthly fee for the service.

Some cable systems operate small studios and/or remote facilities, where low-budget programs (usually sporting events) are produced for subscribers. Most cable systems do not have enough subscribers to interest advertisers, so money for original production must come from the cable system itself. If original programming is demanded by the system's franchise with the city (a "franchise" is a contract with the city to operate the cable system), then the system will usually fulfill the requirement with low-budget activities. Original cable programming is rarely produced for any reason except franchise requirements.

EDUCATION TELEVISION AND VIDEO

Ever since television's invention, educators have been impressed with its potential. Unfortunately, money for quality educational programming has always been tight (television is, and always will be, an advertising medium that is most effective when entertainment programs are used to reach the greatest number of viewers—a fact that makes educational programming difficult). The list of quality educational programs that have been seen by large numbers of viewers is so short as to be embarrassing. PBS's "Sesame Street" and "The Electric Company," both products of the Children's Television Workshop, and some of the programs on "Nickelodeon," a cable network, are among the few educational programs that have received national attention.

In an environment where PBS, the largest

potential distributor of educational programming, must constantly beg its viewers for money, the future of educational programming looks bleak.

Still, many public television stations, particularly those run by universities, do produce many hours of low-budget educational programming. These can provide opportunities for producers and directors who are interested in using television to teach. The market for instructional television programming in the corporate world is growing rapidly as well (see the next section).

There is enormous opportunity in local and regional educational programming, for the right idea, the right package, the right marketing and distribution scheme. Many schools have television studios, and some of these are more lavishly equipped than local stations in the same cities. With videocassette recorders and computers in most elementary, middle, and high schools, and satellite dishes in an increasing number of colleges and universities, new possibilities are emerging. As technology causes some production costs to fall, educational television may become an important component in the production picture.

CORPORATE VIDEO

Nearly all of the Fortune 500 companies use video in training, sales, communications, and meetings. Many own small studios, where staff and/or free-lance producers and directors create relatively simple videotapes for internal use. For the larger productions—presentations for trade shows and the like—outside production companies are usually hired on a free-lance basis.

Besides the traditional meeting, speech or training program, corporations use video in two unusual ways:

Corporate Television Production
This industrial shoot, for Clark Equipment Company, required a combination of office settings and factorylike conditions to show heavy equipment. The office environment was created in a studio with three flats, rented office furniture and art. A single camera, mounted low to the ground (to match the eye level of the executives), covered the action, which was later edited into a finished production. *(Photo courtesy Teletronics)*

1. Interactive Video

With videodiscs and other interactive communications devices, the training marketplace is an area of tremendous opportunity for the producer. It is here that television and computers are most closely allied, where videodiscs are used to permit employees to actually see the results of their decisions. Videodiscs are also used in consumer marketing, especially when new products require more explanation than a salesperson can easily provide. Apple Computer, for example, used an informational videodisc presentation with Dick Cavett to explain word processing, spreadsheets, and other computer applications. Here, the videodisc was used in connection with a touch-sensitive screen (a device that was first developed for use with computers). The disc would display a list of topics on the screen, the viewer would touch the first letter of his or her choice, a small computer would recognize the response and tell the videodisc

player which section to play. This was an extremely simple use of the videodisc; when connected to a computer, a properly designed videodisc program can be more effective than an individual instructor.

2. Teleconferencing

With the development of satellite technology came the realization that short-term rentals of satellite uplinks (which send the television signal up to the satellite), satellite time, and downlinks (which receive the signal) could be a huge time-saver for busy executives. A teleconference usually involves two or more groups of executives in different cities, with each group seated in a television studio (or a conference room with television equipment). The satellite is used to connect the teleconferencing centers, permitting instant sound and

Teleconference
Business meetings can now be held via television, with executives in one city seeing and speaking with their counterparts in another. This television studio was converted into a teleconferencing facility. The monitors show not only words from a TelePrompTer, but the faces of executives at a similar facility some fifteen hundred miles away. Note that the set design allows each executive to see at least one monitor without straining. *(Photo courtesy Reeves Teletape)*

picture interaction. Since there are customarily a few executives in each studio, several cameras are usually involved, and a director cuts between them to create a single feed, just as he or she would do on a conventional television program.

When compared with the cost (both time and money) involved in a business trip, the relatively high cost of teleconferencing can be justified. As the teleconferencing business grows, and more companies realize that this is not a futuristic dream, but a present-day reality, prices will drop, and teleconferences may become as common in the business world as word processors. As the prices of the equipment drop, many companies are building traditional office conference rooms with video capability. In the near future, most office complexes will be hard-wired or connected via microwave to a satellite uplink, which is rented by the hour.

HOME VIDEO

Like pay television, the home video business depends almost entirely upon motion picture product. Since the home video business is dominated by companies owned by movie studios (of the larger "labels," only one company, Vestron Video, operates without a guaranteed supply of releases from a mother studio or production company), most of the titles released by the labels are movies. In an effort to expand their business, most of the labels are releasing nonmovie product, generally as a small part of their overall operation. "Jane Fonda's Workout" has been a tremendous success for the independent Karl Video (now part of Lorimar), and some music programs, like "The Compleat Beatles," have been big sellers. Special products, such as Parker Brothers' VCR-game version of the board game "Clue," have been successful as well.

Most people buy videocassette recorders for "time-shift" viewing. Many rent movies and special programs, like children's shows. Whether a sufficient number of people will buy or rent original productions to justify the costs of production and marketing remains to be seen: the number of successful programs to date is promising, but consumer habits are difficult to predict.

COMMERCIALS

Most television commercials are produced by advertising agencies, who supply a full range of services to their clients, from the conception of the advertising campaign to the creation of the commercials and the buying of air time (i.e., purchasing the time from networks or local stations for running the commercials).

There are a few dozen important agencies, most in New York, Chicago, and Los Angeles, who do the bulk of the commercial work for network and local television, and there are hundreds, perhaps thousands, of smaller agencies. Young & Rubicam (Y&R); J. Walter Thompson (JWT); Ogilvy & Mather; Dancer Fitzgerald Sample; BBDO International; Saatchi & Saatchi/Compton; Doyle Dane Bernbach, and a few others handle most of the work, but smaller agencies, like Ally & Gargano, are also important, especially when a talented client wants something unusual (Ally & Gargano created the award-winning fast-talking Federal Express commercial after seeing a young actor named John Moschitta on "That's Incredible"). These agencies maintain creative staffs, but free-lance producers and directors are used to actually create the commercials (Bob Giraldi, who directed the Michael Jackson "Thriller" video, and Joe Sedelmaier, who directed the Federal Express commercial and Wendy's "Where's the Beef?" spot, are among the best known).

Most network commercials are produced on 35-mm film, which offers excellent color quality and detail, and a traditional "film look." Very few are produced on tape, because decision-makers claim that the "look" of tape is too sharp, too real, and too brash.

Local television stations keep their studio crews busy during the day (e.g., after the morning talk show and before the news) by producing commercials on videotape for department stores, car dealers, and other local merchants. In the larger cities, commercial work keeps both independent studios and directors busy. In the smaller cities, commercials are supplemental to regular activities.

MUSIC VIDEO

Although the glamour of music video production might suggest the reverse, music videos are produced to sell records. When a record company decides to release a single, the decision to produce a music video to promote the song usually follows. Most music videos are produced very quickly—in a matter of a few weeks from conception to delivery—so the relationship between the recording artist and the producer and director is extremely important. Most major acts use one or two music video directors, who understand what the act is about, and how it wants to be portrayed on video. (Incidentally, most music "videos" are produced on film, again because of "the look.")

Videos are usually about four minutes long, usually feature the act in performance (rarely on stage, usually in an exotic location or an abstract situation). Despite budgets of $50,000 to $100,000 or more, videos do not usually earn money for the record companies. MTV does not usually pay the record companies for use of these "promotional materials" (al-though it does pay for the exclusive right to premiere some of the best new videos); other music video outlets also consider the clips to be promotional materials, and do not pay to play them. The record companies produce the videos for one reason: to sell records. If the video can later be released to home video, or otherwise packaged to make some money, the income usually comes as a pleasant surprise. Michael Jackson's "Thriller" video, for example, was very successful when it was repackaged as part of "The Making of Michael Jackson's 'Thriller,'" which was bought by Showtime and by several hundred thousand VCR owners.

OTHER MARKETS

With new developments like home video cameras and comparatively inexpensive professional recorders ($35,000 now buys a broadcast-quality camera/recorder unit), the market for talented producers and directors is expanding into new areas. One videodisc producer, for example, was hired by a city agency to record a video picture of every house in Rochester, New York, on disc (a single videodisc can store a few hundred thousand images). Some institutions are now offering grants to video artists, who use technology to stretch the limitations of the medium. Many of the major league baseball parks have installed giant screens in addition to (or in place of) scoreboards, which show player close-ups and statistics, key plays, and promotional information. The production of the show on the scoreboard requires a producer and a director, as well as a full crew. As video and computer technology continue to develop, television producers and directors who understand the changes will see many new markets and new opportunities for their talents.

TWO

Working in Production

Television production, when done properly, is a cooperative form. The best programs are the result of several creative contributors: writers, producers, directors, performers, and any number of other people behind the scenes. Like the movies, however, the cooperation takes the form of a benevolent dictatorship. In the movies, the director is in charge. In television, the person in charge is most often the producer.

The producer is the boss, the prime creative force, the keeper of the budget, and the administrator. The director is the craftsman, the one who must ultimately transform script, talent, and ideas into a completed film or videotape. Members of the production staff are responsible for the detail work. Whatever his or her role, each of these people is attached to a television project to improve the likelihood of its success. Television is a game of popularity. In commercial television, pay television, and now cable television, popularity is measured in ratings; in home video, it is measured in sales; in business video, in either sales (if the product is sold to corporations), or the effectiveness of the presentation (if the product is used internally). The success of a project can oftentimes be linked to the skill of producer and director, to their enthusiasm, to their desire to make the project a success, to their constant striving for the best possible scripts, the best possible performances, for interesting and unusual treatments, for engaging sets and enticing music, for a desire to keep the entire presentation fresh and appealing. There is a special kind of dedication required of producers and directors, and of everyone else involved with a television production. In a word, this dedication is what being a "professional" is all about.

PRODUCERS

The word "producer" is most often associated with the movies, probably because movie producers are starmakers, and because they've always gotten a lot of press. Theatrical producers are also well known, particularly among those with money to invest; their primary function is in the packaging of the project (putting the right people with the right project), and in securing the financing. A television producer fulfills a totally different role.

Television has been called a "producer's medium," because the producer's vision provides the guidance and direction for every project (by comparison, the movies are a "director's medium," because it is the director's vision that establishes the style of the project).

Anyone with the word "producer" in his or her title has a single responsibility: to achieve and maintain a level of excellence, while completing each project on time and on budget. The producer is ultimately responsible for everything that appears on the screen, every sound, every cue, every word in the script. Every decision that a producer makes has a definite effect on the finished product: the choice of the best possible staff, the decision to stay late and rework a scene, the last-minute instinct that two hosts would be better than one, the insistence that the set pieces be repainted a shade darker, the fight for an extra hour of studio time, *everything* that affects the finished product. And, in the end, the finished product is the only thing that matters. Good producers know that they are judged by what appears on the screen, not on the number of friends (or enemies) made in the course of production. And savvy producers know how to get everything they need—and still manage to be rehired despite the occasional indiscretion.

1. Executive Producer

In local television, the executive producer is usually the program director (PD), or a person assigned by the PD to supervise production. In network television, the executive producer is usually the packager, or someone assigned to supervise. On home video productions, the executive producer may be the packager, the president of the home video company, an advertiser, a distributor, or one of the key dealmakers. There are many different situations, because television productions are collaborative efforts, and because many productions come about as a result of high-level executive decisions. Make no mistake about it, though: the executive producer is always the boss, the one who makes the final decisions.

2. Executive in Charge of Production

This is a term used in big-budget projects, where the executive producer does not have the time (or, sometimes, the expertise) to supervise the entire production. Many executive producers have a gift for finding money, for selling projects, for developing successful shows, for building superior staffs of production people, but require an executive in charge of production to run the operation. The executive in charge of production, or production executive (a similar, but not always identical, term), is usually associated with the bigger projects; for smaller projects, the executive producer generally depends directly upon the producer, or upon a combination of the producer and the production manager.

3. Supervising Producer

When several people have been given producing titles, a prefix like "supervising," "senior," "coordinating" or "co-" may be used, generally in accordance with the company traditions or executive privilege. "Pecking orders" may vary.

On a network series, the supervising producer is usually responsible for the entire series, with one producer assigned to each episode.

4. Producer

In the good old days, every show had one producer, who took charge of the entire operation. Now that there are supervising producers, senior producers, senior executive producers, and coordinating producers, it's hard to know who does what. The person who takes charge of the day-to-day operation is now called either the line producer (the producer who has direct line responsibility), the supervising producer, or, simply, the producer. On a production where much of the responsibility lies in the planning, a studio producer may be hired to supervise only the actual studio activities, while other producers prepare future shows. The morning shows at the networks use this system, rotating studio responsibilities on a regular basis.

The important thing here is flexibility: the line producer on one program may be the supervising producer on another and a senior associate producer on another.

5. Field Producer

With the popularity of portable video cameras and tape recorders came the television magazine format, whose programs are generally staffed by hybrid producer-directors called field producers. The job combines producing and directing activities: the field producer plans the shoot, directs the camera operator and talent while on location, and supervises the editor and the postproduction process.

6. Associate Producer

Always second to a producer, the associate producer's job description depends entirely upon the specific requirements of the production and the needs of the producer. The AP must be extremely loyal, must present a united front with the producer regardless of the situation (within reason, of course). A good AP will be able to take on almost any of the producer's responsibilities in a pinch. Associate producers oftentimes play an important role in "smoothing over" relationships, making certain that a producer or director's occasional harsh words or rushed treatment do not affect the success of the project.

On a talk show or a magazine program, an associate producer is usually responsible for arranging guest appearances and writing the interviews (the process of convincing a guest to appear is called "booking," so APs and others who do this are called "bookers"). Most talk shows employ several APs, each one responsible for a certain number of segments per show or per week. (On network talk shows, the big bookings are left to segment producers, who combine booking and writing expertise with their experience as producers.)

On most network, pay TV and syndicated specials, variety shows and sitcoms, the associate producer is in charge of all production logistics: arrangement of the production office, hiring of lesser staff members, schedules, crews, facilities, script distribution, contract administration, union payments, and just about every other detail that the producer is too busy to think about. If the budget permits, the associate producer is supported by a production manager (see below) and a unit manager (primarily a budget watchdog and facilities expediter, commonplace on network shows).

The function of the associate producer is, ultimately, to lighten the load on the producer and director, so that they may concern themselves only with the important things, like the script, the performers, the pace, and new production elements that will add a few rating points. A good associate producer makes the wildest dreams of the producer or director come true, usually with a single phone call:

Producer: We'll need three helicopters, sixteen redheads who can water-ski, and seventeen golden retrievers.

Director: In an hour.

Associate Producer: No problem.

On a network drama, the associate producer may be the detail person, and may also be the staff member responsible for editing the program. (The more senior version of this job is post production supervisor.)

The title of associate producer is, occasionally, used by a producer as a perk. It may be given to a director who needs producer credit, to a writer who is working on a nonunion show and cannot accept the writer's credit (not to be encouraged, but it happens), and it may be given to a financer (this is typical in the movies). Many field producers are called associate producers, because their jobs are similar to those performed by talk show APs—they are responsible for segments.

7. Production Manager

The term comes from movies and film production, where the production manager is responsible for budgets and union personnel. The production manager is the person with both feet on the floor, the trustworthy, knowledgeable soul who tactfully knows how to tell a producer that he or she is over budget, who quietly handles minor problems with union members to prevent the filing of grievances. Like the associate producer, the production manager takes care of details. But the PM's details are more likely to encompass the unglamorous: the sets, the videotapes, the accountants, while the AP handles the potential crazies: the talent, the writers, the producers, the guests. A good production manager, usually working closely with a good associate producer, is an important force in the completion of a production on time, on budget, and with a minimum of serious problems.

8. Unit Manager

Primarily a network term, the job combines production management with production supervision for the network. Every network employs a staff of unit managers who work with independent production companies and with the network's internal news and sports divisions, making certain that all of the network departments are working together. The unit manager takes care of scheduling network facilities, keeping the director appraised of crew changes, arranging for the studio and stage crew for the loading and striking of the set, setting up editing time, monitoring the budget, and making certain that outside vendors are paid (e.g., prop rental houses). The unit manager also represents the network's interest on the technical and logistical sides of the production.

When the networks rent their studios for projects outside the network's schedules (this is common for syndicated game shows in Los Angeles, for example), a unit manager makes certain that the facilities are working properly, but operates in a service, rather than a supervising, capacity.

9. Production Assistant

Just about every producer or director has spent some time as a production assistant (PA). Most often an entry-level job, there are usually several PAs on all but the smallest production projects. The responsibilities of the PA are usually set by the associate producer, who must manage all of the production details. The PAs are used to take care of those details, typing scripts, spending hours at a duplicating machine, running for coffee and bagels, finding those golden retrievers, and so forth. A good associate producer will organize the production staff (the PAs as a group are oftentimes called the production staff) with one set of responsibilities in the office, a second set in the studio, and, in some cases, a third set in the editing room.

PAs are usually given a list of responsibilities, checklists prepared by the associate producer. When I worked as a PA on a game show, I spent most of my time in the office typing questions (written by the show's staff of writers), and occasionally researching a specific question for one of the writers. In the studio, before the show, I was responsible for the large cards that fit into the gameboard, making sure that the spellings were correct and that the cards were inserted into the board in their proper positions. I was also responsible for the contestant's nameplates. During the taping, I stood behind the scoreboard operator and told him which score to enter for each contestant. After the show, I gathered all of the questions and filed them for future use. Another of the PAs was responsible for all of the prize cards, and the announcer copy that accompanied them, the forms that were to be signed by the contestants, and the cue cards (making sure that they were all in order). She also sat in on editing sessions when an extra hand was needed. Another did nothing but answer viewer mail (and, when that was slow, lent a hand to the contestant coordinator, grading the written tests taken by every applicant). A production secretary just typed questions all day. All four of us were aggressive (and, in fact, all four "made it"—one is a producer at "Good Morning America," another has been an associate director for several prime time series, another works at ABC News).

A good PA with plans for advancement knows one key phrase: "I'll take care of it."

The job is usually a stepping-stone to associate producer or associate director; most successful producers and directors started their careers as PAs.

On some videotape programs, when the production assistant works in the control room with a stopwatch (usually a third hand for the associate director), the job is under the jurisdiction of the Directors Guild of America (DGA).

In most situations, producers pay PAs as little as possible (less than a secretary, if you're a newcomer). Many producers like to think of the low pay as an incentive: if you're willing to work very long hours for very little money and even less encouragement, you may have what it takes to make it in television.

Incidentally, there are some special titles, like production coordinator, awarded to production assistants who have not yet earned the associate producer ranking. Fortunately, these titles come with a bit more money and responsibility. Progress may be discouragingly slow in the early stages, but it is important to remember that a good PA will be noticed by the rest of the staff, and that these early friendships may become important contacts in your later career.

DIRECTORS

Unlike a movie director, who exercises absolute control over all that is seen and heard in a motion picture, and unlike a theatrical director, who is permitted a certain degree of temperament, the television director is expected to be the consummate professional, keeping a careful eye on the clock as he or she creates the best possible product while working under limitations of budget and creative range determined by the producer.

Throughout the development and preproduction phases, the producer is in charge. A few days before the shooting begins, the director takes charge. Ideally, the producer becomes a trusted adviser and troubleshooter, with the director and other production personnel now ready to execute all that the producer and staff have planned. In practice, the director looks to the producer for a sense of control, and for approvals over changes in the original plan. After working together for some time, a producer-and-director team will develop instinctual feelings for what "works" for the other. Frequently, an associate producer or associate director is important in this process as well.

Every director is expected to be a master of three areas of production: staging (guiding the performers for maximum impact, moving the cameras to cover the action, and using the set and the lighting to its maximum advantage), pacing (working in the control room, cutting between cameras, instinctively knowing when a music cue should hit, when to roll tape, when to give the performer a speed-up or a stretch), and editing (creating individual pieces in the studio or in the field that will fit together to tell the story with maximum effect). Some directors are better at certain skills (e.g., working with the performers) than others (composing shots). The best directors make it all look so easy; the results of intense concentration and years of practice are immediately apparent when watching the work of a true professional. With some innate ability and some training, most people who want to direct can cut a show reasonably well. There is, however, a difference between directing as a skill and directing as an art.

1. Director

A director may become involved in a project at its inception, or at any stage prior to the actual on-camera production. Most directors, aside from local news directors, are free-lancers, so they schedule projects sequentially. Typically, the director sits in on some of the planning meetings (particularly when set design, number and location of cameras and microphones, schedules, and logistics are being discussed). The director usually consults during the preproduction phases, or

takes charge of certain production areas entirely (e.g., special effects). The director may become a full-time staff member only a few days before shooting begins. Some directors insist on being involved in the editing process, while others prefer to finish shooting and then move on to another job.

2. ISO and Second Unit Directors

Most productions require the services of only one director, who supervises the crew, works with the talent, and "cuts" the show (e.g., cutting between several different cameras, cuing, etc.). More than one director is required when a large event, like a political convention or a concert, is being covered by a large number of cameras (six or more). In these situations, one or more directors cut between smaller groups of three or four cameras, and feed only one signal to the supervising director in a mobile van or in a home-base control room located at the main studio facility. Since these cameras are isolated from the main operation, the feed is called an ISO feed and the director is called an ISO director. This parallels the work of a second unit director, who works with one crew shooting background footage and sometimes stunts and other specialties, saving both director in charge and the talent hours of work. Second unit work is common on motion pictures when secondary scenes are important to telling the story; second unit work is not as common in television, because key performers appear in almost every scene.

Although television directors spend most of their time in the control room directing cameras, they also work with talent. The directors of network sitcoms, dramas, and soap operas (more politely called "daytime dramas") work closely with performers, developing characters and motivations, blocking the action, operating in the style of motion picture directors (under considerably greater time pressures, however). Directors of business and educational programs work with inexperienced talent (like corporate executives, teachers, industry experts, etc.). Directors of news and talk shows offer occasional comments and suggestions, but are not usually charged with the quality of the performances (this being the producer's responsibility).

The Directors Guild of America is the trade union for all motion picture directors, and all directors working in the following areas of television: broadcast networks, local stations in large markets, and national television commercials. Most of the other markets (cable television and home video, for example) are not under DGA jurisdiction. In theory, this means that only nonunion directors can work in these markets. In practice, the DGA is in the process of redefining its rules to permit its directors—most are far more skilled than their nonunion counterparts—to work in the new media as well.

3. Associate Director

More than just the director's assistant, the AD is an essential link between the director and the crew, the one who makes certain that the logistics of the studio, the control room, and the entire facility work together as one. Since the AD works for the director, the amount of actual directing involved in the job does vary with each director and each situation. Some directors demand that their ADs take an active role, setting up shots, cuing, and supervising the studio crew. Others use the AD as a guy/gal Friday, requiring accurate note-taking, coordination of mundane tasks like the labeling of recorded videotapes, and so forth. Some directors find it useful to have the AD handle rehearsals from the control room, so they may be on the studio floor with the talent, thus providing the AD with an opportunity to direct. Most directors require their ADs to be adept

in the use of a stopwatch and in "back-timing" (counting the number of minutes and seconds remaining in the program). In any case, a good AD will work closely with the crew, usually accepting little credit for a considerable amount of work. The director, whose production is likely to move more smoothly with a top-notch AD, gets the credit.

The AD job is frequently a stepping-stone to director, especially when an AD is given the chance to direct rehearsals, and is trained by a good director. It is not unusual to find an AD doing the job of second unit director, stage manager, associate producer, or field producer because all involve functions that can be helpful in becoming a director.

In some instances, associate directors are called assistant directors. To be accurate, associate director is the videotape term, and assistant director is the film term. Don't worry too much if the terms are used interchangeably; the job is the same in any case.

4. Stage Manager

The concept of stage manager is borrowed from the theater, where the SM is the person who runs the day-to-day production after the director has completed the initial staging. In television, the stage manager is the director's voice on the studio floor, connected to the director in the control room via headset. As a production is being rehearsed, the SM is busy coordinating studio elements, keeping an eye on the whole studio including the stagehands (there may be as many as a dozen on a larger production) and solving problems for the director. During the production, the SM is responsible mainly for the performers (moving them on and off stage at the appropriate times) and for cuing—using hand signals to tell the performer when to begin, how much time is remaining, when to slow down, when to speed up, and when to "stretch" (a "stretch" is required when three minutes of show time remain with only two minutes of script—it is an opportunity for a performer to show skills as an ad-libber). The stage manager is clearly a supervisory person, responsible for the scenery, the props, sometimes the cue cards. The job of the SM, which is essentially a coordination function in the studio, complements the job of the AD, who coordinates the technical elements of the production. (In fact, the DGA, under whose jurisdiction SMs work in many markets, offer identical pay scales for stage managers and associate directors.)

Some local stations see the job of stage manager as a function that can be performed only by a technician (usually the result of a contract signed many years before, or a station tradition). Small productions limit the job of the SM to cuing, and hire either a production assistant or a college intern to do the work.

ON BEING A PROFESSIONAL

Television production is a "people business." The work oftentimes demands personal sacrifices, long hours, and, above all, patience. On a successful project, the performers, the production staff, and the technical and stage crews will work together as a family. Teamwork is not just a catchphrase, it is a very real tool that can save hours of studio time and thousands of dollars. Teamwork helps a production to take on some "magic" as well. The results of teamwork are frequently evident in the pace, the style, and the overall effectiveness of the finished production.

One of the greatest compliments that can be paid to a producer, director, performer, stagehand, secretary, or technician is that he or she is "a real professional." People get jobs because they are professionals. People don't get jobs if they complain, if they cause problems on the set, if they put their personal lives ahead of the production, if they act in an "un-professional" manner. All of this goes back to a "show must go on" mentality that is a very real part of almost every production—television projects seem to develop a life of their own, more demanding than any child or spouse could ever be. Night after night, when the script's not exactly right, when a guest suddenly drops out, when there are a few small details that must be resolved before the next step can begin, the stress can start to show. It's vitally important to one's career to BE THERE whenever a situation arises, to place the production in a more important position than birthdays, anniversaries, prepaid vacations, job interviews, and other occasions that might seem important in "real life." Only a birth or a death is considered a good reason to be away at a crucial time—everything else must be rescheduled to fall between the busy times (and, please, make sure they *are* rescheduled—a bad balance of creative work and personal priorities has already destroyed too many relationships).

Professional, and unprofessional, behavior can set the style for the entire production. A director who, for example, returns from a lunch break late (a mortal sin) will prompt similar behavior from other people on the set ("We never start right on time anyway"). (Time really is money in the studio—the hourly rates for studio and crew keep registering whether anything is being accomplished or not.) A producer who is perceived as uncaring will eventually generate resentment, a poisonous commodity in any creative endeavor. An on-set argument with an important performer can set everyone on edge for the rest of the day; ripple effects can be seen in the pace of events, the performer's sense of timing and overall effectiveness, as well as the amount of time wasted on normally simple tasks.

A professional remains calm and projects a feeling that everything is under control. A professional treats everyone connected with the project with respect—not just the key performers and a few favorite staff members. A professional anticipates problems, and deals with them before they become disasters. And, trite as it may seem, it really is important to thank everyone even remotely connected with the project—if it's sincere, it really is appreciated. A professional is a teacher, a business-person, a psychiatrist who knows how to handle tempers, tension, egos, accidents. She's the extra hand that comes from nowhere because a prop fell out of the performer's reach. He's the PA who knows, instinctively, that the kid next to the host is going to start crying and makes him laugh instead. He's the one who makes the right quiet comment that breaks the tension, and permits the show to go on. She's the one who arrives first, leaves last. And he's also the one who manages to pack up early on a summer Friday, because the week's work is done (and dismisses the whole staff early as well).

When tension strikes (and it will, no matter how large or how small the project, no matter how well or how poorly prepared), it is important to remember that the key to success, regardless of the situation, is to play the role of the professional. Once you've played the role a few times, it will come naturally. And you'll never work any other way again.

THREE

Development

Television production costs money. With few exceptions, producers and directors work with "OPM"—a tongue-in-cheek phrase meaning "other people's money." This chapter is about the most important aspect of creative life: how to get other people to give you their money so that you can spend it.

The term "development" refers to the period before production begins. This period may be a few days, is usually a few weeks or months, and may even be years on the biggest projects. The development process usually begins when somebody says, "I have a good idea," and somebody else agrees. It continues as the project is described on paper as a script, a treatment, or a storyboard, through meetings, revisions, budgets, demonstration tapes and pilots. Eventually, a decision is made. The decision is most often "No," sometimes "Make a few changes and we'll discuss it," and, occasionally, "Yes." One yes after months of nos makes it all worthwhile.

Creative people are full of ideas, and there are channels for the submission of these ideas in almost every television station, home video company, and so forth. The companies themselves also generate ideas, which are devel-

oped and produced by either hiring free-lance producers and directors, or by working with staff personnel. In either case, the development of production projects is always a creative enterprise, with some good business reasoning behind it.

Programming executives are most often the people who groom ideas for approval. All of the studios maintain program development staffs for the development of network and syndication projects. Many of the PBS stations develop both programming and program financing concepts at the program executive level as well. At the commercial and pay television networks, the program development executives are constantly reviewing new project submissions (with those submitted by established entities taken most seriously).

Local stations and cable systems don't do very much development because they don't do very much original production. In these cases, sponsorship is usually the only phase that requires development, and this is the job of the sales department.

Corporate video development is rarely a matter of "I have an idea" at the producer/ director level. Projects are usually requested by department heads, and then facilitated by the production department at the corporation. There are many instances, however, where clever staff people have spotted a specific company problem and created a videotape to solve it. Generally, projects are not generated by the creative staff, but by the management staffs instead.

Commercial development is done by the creative people at the advertising agency, who win the right to produce commercials for a specific account. Sometimes commercial concepts are developed with outside people (especially if a complicated animation or a stunt is to be part of the finished product, or if a celebrity is to be involved), but most of the commercials are developed in-house, and then produced by free-lancers.

CONCEPTS

A concept is a raw idea in need of further development. It is on its way to becoming a real project when it takes on structure, when the basic idea can be visualized, at least in rough form, as a completed production. As the concept takes shape, the creator should find some time to either write it down (writing forces concentration, so the end product is likely to be more complete than anticipated) or to talk it through with someone (just be sure that the someone will not adopt it as their own, or assume a partnership). Concepts are "a dime a dozen." If a concept is not treated to some development, it will never grow. Development reveals the quality of the idea and provides the justification for further time and energy.

PROTECTION FOR IDEAS

Once again, concepts are best shared only with the closest of friends and associates; a

good idea, especially one in raw form, can easily become the basis for somebody else's project if overheard at a party or in casual office conversation. The law offers little protection for ideas—only finished projects can be protected under copyright.

Why? Because there are very few ideas that are truly new. Most ideas that seem to be new are variations on existing ideas, with fresh faces, with some window dressing. Ideas for television projects are constantly being discussed in hundreds of offices, so many ideas, in fact, that program development executives sometimes complain that they run together. What may seem to be a new idea in today's meeting may have been submitted the week before, and forgotten.

Creative people who submit original ideas have little protection. Some attorneys suggest that the concept, treatment, script, whatever be mailed *via registered mail* to the creator's address, and left *unopened*. Then if need be, a sealed envelope with a verified date (verified by a government agency: the post office) can be submitted to a judge. Other attorneys consider this to be utter nonsense, and simply recommend a trail of correspondence between creator and programming company, which they admit to be almost as useless. Proving that an idea was "stolen" is very nearly impossible. For this reason, some companies are using "submission agreements," which say, essentially, that the company is constantly receiving ideas and that your idea may be identical to one that they have received or may receive in the future—so you have no recourse if they use an idea identical to your idea, even if they make a million dollars with it.

How do you protect an idea? You don't, really. You protect a relationship instead. Dealing with reputable people, preferably through an agent if you're working in the entertainment world, cuts down the chances of being ripped off. If your idea is abducted, learn from the mistake, and move on. Lawsuits can be damaging to reputations, regardless of who wins.

Just be careful, don't talk too much, and, when someone steals your idea . . . think of another one and move on.

TREATMENTS

A treatment is a two- to ten-page story, a synopsis of a full script. The process of writing a treatment forces the writer to lay out the entire plot line, while fleshing out individual scenes and characters. Some treatments include a little dialogue as detail, but most are written entirely in prose. A treatment may be organized as a short story, with one scene flowing into the next, or as an outline, with each scene described separately.

A treatment is a tool that is useful to both the writer and the program executive. Reading a treatment provides a knowledgeable reader with information that can be used to start thinking about casting, budgeting, and packaging. Most important of all, a treatment is the first step toward a finished script.

In the entertainment world, treatments, scripts, and their revisions are the basis for "step deals." Here's how a step deal works:

In an initial meeting, the creative person suggests an idea to a development executive. If the creative person has credentials, the development executive will negotiate a contract with a series of steps: the first draft treatment, the second draft treatment, the polished treatment, the first draft script, the second draft script, the third draft script, the polish, the pilot script . . . and so on, with some variations, through production. Each of these steps is keyed to a progressively higher dollar figure. From the creator's point of view, draft treatments don't pay much, but they are the first step to potentially greater returns. From the development executive's perspective, the dollar commitments are relatively low in the formative stages, and appropriately higher as the project gains momentum and takes shape as a salable property.

If the creative person has no credentials, or is just starting out as a writer, treatments and even scripts are frequently written on speculation ("on spec"—pronounced "speck"). Since a writer can only be judged on his or her work, most writers start this way. Although writing for yourself doesn't help pay the rent, writing sample scripts forces you to structure your thoughts and to develop good working habits, and, to be honest, these scripts will probably do nothing more than get you a few

Sample Treatment
"Unfriendly Neighbors"   By Howard J. Blumenthal

It all starts in the lot behind the schoolyard. After a month of chronic headaches, several third-graders notice smoke coming from the ground. When Michael refuses to go to school one morning, his parents decide to investigate. Hours later, their worst thoughts are confirmed. Toxic wastes.

Setting the cleanup effort in motion, however, is not a simple matter. While the school is located in Bradford, the lot is actually in the neighboring town of Amity. And everybody knows that people in Bradford don't talk to the people in Amity. And vice versa.

When a legal loophole causes delay, Michael's parents and the people of Bradford take the matter into their own hands, provoking action from the people of Amity (particularly old Al Hansen, whose factory dumped the wastes in the first place). In time, the intensity of the skirmishes mount, and a child is seriously injured. The national media take an interest, and stories about "the Amity-Bradford war" fuel even stronger feelings amongst the townspeople. Michael's father organizes a vigilante force; they will force Hansen to clean up the waste.

On the night that the Bradford vigilantes raid Amity . . .

[And so on.]

meetings with agents and producers who can help you get other assignments. Many producers working in prime time television started as writers (the others started as program development executives), so the endless hours of writing without any hope of seeing your work produced are best considered an investment in the future. One of the producers of "Hill Street Blues" started this way—picking up odd jobs, and continuously writing scripts that nobody produced. Needless to say, the effort, which ran over the course of several years, was worth it. The process can be tedious and nerve-racking, but that's what this business is all about: perseverance.

PROPOSALS

When a creator puts some initial thoughts on paper, it is in one of two forms. If the project is to be a fiction piece, like a motion picture, a comedy, or a dramatic program, a treatment will tell the story, in prose, in the course of three to ten pages. (You'll find more about treatments in the next section.) If the project is nonfiction, like a home video exercise program or an idea for a local television fashion show sponsored by a department store, the idea is best expressed in a proposal.

A proposal may be as brief as a few pages or as long as a small book, depending, of course,

Program Proposal

A good program proposal answers the following questions, in a detailed, yet easy-to-read format:

1. What's the basic *concept*? How or why will the project be successful? What's the market, what's the competition?
2. How does the *format* work? Will this project be effective as a half-hour, an hour, or, in the case of a music video, under five minutes?
3. Are there any special production considerations?

Anything special about the idea that will drive the budget up (or keep the budget way down)?
4. What's the *package*? Is there a star involved, or any other "hooks" that are likely to provoke audience interest?

The sample proposal, on a project called *Healthy, Wealthy, and Wise,* includes much of the information in a good, straightforward presentation. Note that in actual program proposals, the various topics (i.e., "Concept," "Appeal," etc.) would appear on individual typewritten pages.

Healthy, Wealthy, and Wise

A VCA Program for
Cable Health Network

Contact: Howard J. Blumenthal, General Manager, VCA Programs, Inc.

Concept
There are many ways to disseminate information. One of the most effective means of getting a message across involves an easy trick, known to schoolteachers for years and years: Make Learning Fun.

"Healthy, Wealthy, and Wise" is a learning game, a television game designed to entertain while teaching the basics of preventive medicine,

nutrition, exercise, and the tenets of a healthy life-style.

Appeal
This is a family program, designed to entertain and inform. Every question will present useful information, in a simple and helpful fashion.

The program will be produced in the classic tradition of game shows, in the style of "Password" and

"Jeopardy!" The game will be exciting, but the information will not be eclipsed by screaming contestants and flashing lights. The game will be fun to watch, fun to play, and always full of worthwhile information.

Questions as Information
The use of a game format in a program about health information requires careful research and

verification by experts. Under normal circumstances, a "quiz show" question is "double-sourced"; that is, every question must be verified by two reliable and verifiable sources. Here, a third step is required: every question must be approved by a distinguished medical authority. There can be no gray areas, no dubious right and wrong answers. Every question must have one right answer, and only one right answer.

Within the structure of the game, correct answers, which contain correct information, will be audibly and visually reinforced. Wrong answers will always be identified as such, and correct answers given. This is a very important point: through all of the entertainment and fun, we must never lose sight of our responsibility to the audience.

A "judge" will be present at all videotapings, to verify correct answers and to preside over questionable answers. He or she will appear off camera (but may appear on camera when called upon).

Beginning the Game

Three family teams compete. Each family has three members: a mother, a father, and a child (over eight years old). A fourth team member, also a child, plays a "Computer Bonus" game, backstage, which will be discussed later.

Each of the family members is dressed in exercise clothes (e.g., a jogging suit); all family members will be required to be physical at some time during the show.

Play begins with a question for all three families:

"This question is to be answered only by the kids: 'How many situps can Dad do in fifteen seconds?' "

The kids write down their

answers, and do not show them to anyone. The dads compete, while the families and the studio audience cheer them on. We then see the kids' answers—the family with the closest answer wins control of the gameboard and the right to answer questions.

The Gameboard

The gameboard contains seven categories, each with eight questions. The value of each question is keyed to a portion of the CHN logo.

On our sample gameboard, the category OLD WIVES' TALES would contain questions about folk medicine. PETS would contain all sorts of questions about having animals as pets (both medical and nonmedical). VITAMINS AND MINERALS is self-explanatory. HAVE A HEART contains questions about your heart and circulatory system. LIFESAVERS is all about first aid, THE BODYWORKS contains questions about how your body works, and READ THE LABEL is all about nutritional information that can be found on food labels.

[On the next page of the actual proposal is a diagram of the gameboard in color, followed by an explanation of how to play the game, with sample categories, sample questions and answers.]

About VCA Programs, Inc.

VCA Programs, Inc., has produced two award-winning productions to date: "The Boys of Summer," based on the best-selling book by Roger Kahn, and "The World's Greatest Photography Course." Each is a co-production.

Howard J. Blumenthal is the general manager of VCA Programs,

Inc., and the creator of this program. Howard was in charge of all interactive game development for QUBE, and the producer of several successful game show formats, including QUBE's very first interactive production, "How Do You Like Your Eggs?," starring Bill Cullen. Howard is well known in the game show world as a writer and as a producer. He has been involved with game shows for all three networks.

Notes

1. This program will be produced in New York City, in front of a live audience. (Several programs will be produced on each studio day.)
2. All contestants will be subject to eligibility requirements: they will be tested to verify their basic knowledge of health and related fields, and they will be interviewed to determine their ability to perform on camera. Family units need not be actual families (but this rule will not be publicized).
3. Prizes will be provided by the manufacturers, or by a firm whose business is providing prizes in exchange for standard game show promotion.
4. All three families playing in the studio will be playing for three families at home as well. These match-ups will be determined as a result of a random drawing of postcards (mailed to the show by home viewers) immediately prior to the taping. Please note that there may be some legal questions about lottery law here, which should be discussed prior to the start of production.

on the complexity of the project. A good proposal will entice the reader in the first few pages, then explain the concept in detail, and finally offer some marketing or business rationale for the project (demographics, star value, revenue projects, publicity tie-ins, etc.). Biographies of the key players should be supplied if the proposal is written for someone who is unfamiliar with their backgrounds. Inclusion of budgets in preliminary proposals is unnecessary, except in special situations. A premature budget that is too high or too low can cause a project to lose momentum in the early stages.

Since proposals generally are reviewed by a few decision-makers, they should sell even when the creator cannot be present. Graphics, particularly on the cover, invite interest (but they have to be good—a bad graphic can kill a project). An unusual style of presentation can make the difference, provided it is appropriate to the situation. Since television is a visual medium, proposals can suggest the visual style of the finished product. The logical extension of this presentation concept is the storyboard, which is described in a subsequent section in this chapter.

GETTING STARTED AS A WRITER, WRITER-PRODUCER, OR WRITER-DIRECTOR

If you're going to try to make it in prime time television as a writer, writer-director, or writer-producer, develop your basic writing skills at home and then move to the Los Angeles area as soon as you have the confidence (most people move without the money, you won't be the first). You will learn more in the first month—once you've managed to meet a few people by calling your best friend's cousin's old girlfriend who works as a receptionist at Paramount—than you could ever learn by writing in the attic in Braintree, Massachusetts or Upper Arlington, Ohio. Take advantage of any schooling that is available, get the college education (a liberal arts background really does provide a foundation), work closely with any professors or fellow students who can help to critique your work, and then MOVE! You will never make it as a television writer or producer or director—assuming entertainment is your game—by remaining in Seattle or Atlanta and mailing your submissions from there. It's a contact game, and you've got to be there to play.

SCRIPTS

Though few outsiders realize it, every television production—even talk shows, game shows, and sporting events—has some sort of script. A script is an important tool, not only because it provides performers with the words to say, but because it provides a definite structure for every second of the production. The production manager works out budgets based on a final script. The director uses the script to make casting decisions, to plot camera angles, to plan the daily or hourly production schedule, and to lay out the final edit. Program executives use the script to show their superiors what the project is all about. With all of these people depending upon the finished script, a great deal of attention is paid to developing a script that will "work." A great deal of time is spent in the script development process, whether on a weekly series, where many writers submit scripts to the producers, who choose the best ones, or on a made-for-television movie, where several years of development and packaging may be required to bring a project though to production. Changes are the rule, not the exception, throughout the script development period, because script changes are cheaper than on-set or postproduction changes, and because a final script is required before anyone will give the okay for production to begin.

A script that tells a story is usually the most difficult to write (the writer must create a half-hour or hour-long play under the pressure of a deadline) and the most likely to see dramatic changes before production begins (because performers and directors read the entire script aloud several times before production begins, so the finished script is really something of a collaboration). Scripts used in comedies, dramas, soap operas, and movies (including made-for-TV movies) concentrate on dialogue, and provide the director with limited information about the camera's point of view, and about cutting the show. Each scene is individually numbered, because shooting is usually done out of sequence ("While we're in the junkyard, we'll shoot scenes 9, 21, 37, and 39; we'll do 2, 5, 11, 16, and 30 at the school tomorrow"; or "Julie has that movie next week, so we'll have to shoot all of her scenes, numbers 2, 7, 12, 34, and 39 before Monday").

Scripts written for nonentertainment programs are not subject to the same kind of scrutiny, except, perhaps, in the world of corporate video, where executives may require considerable input in the script and in other production decisions.

Scripts at local stations are usually news scripts, which are written very quickly, under considerable pressure, and are carefully worded to conform to the style of the anchorperson (who should do little more than change a few words, unless, of course, the anchor acts as a "managing editor," which is the case on some network news programs). News scripts are usually typed for the TelePrompTer (a device that projects the script, in sync with the performer's reading speed, in front of the camera lens), with a single column of copy centered, in large type. This gives the director plenty of room for making notes. Handwritten changes, even seconds before air, are easy to

make, because the prompter script is really just a piece of typed paper moving on a conveyor belt under an inexpensive television camera. Many late-breaking stories have been written from a producer's conversation on the control room phone, and placed under the prompter camera only a few seconds later ("This just in . . ."). Local news scripts, incidentally, are frequently typed onto carbon packs, to avoid the cost (and time) of copying by machine.

"Textbook-style" television scripts, with the audio column neatly typed on the right (with dialogue, narration, music, and effects information) and the video column on the left (with all picture information), are not used as often as they should be. Convenience is the rule, with tradition running a close second. Still, the split-column form is indispensable for complicated productions, particularly those that are live, or recorded live on tape, and especially when the director must follow not only the performers, but integrate graphics, music, and effects as well. The big problem with this script format is the amount of time it takes to type, retype, and revise both of the columns—problems that will be resolved with more sophisticated word processing programs, no doubt. Split-column scripts are most common when there is enough time to plan the entire production properly, when the writer can really take the time to create not only the words, but the images as well. This script format is common on magazine-style productions, where visuals tell most of the story (narration is secondary in a well-produced field piece), in instructional programs (where time and deadlines are secondary to educational value), and special types of programs like game shows, where a detailed script is essential in the early stages and then very nearly discarded as everyone learns the routine (watch any game show and take note of the number of sound effects, music cues, set changes, off-stage announcer cues, graphics,

Sample Script Pages

The first four pages from an imaginary talk show entitled "P.M. America," showing video information on the left, and audio on the right.

The script is straightforward, but may require some decoded abbreviations.

VT indicates a videotape playback.

EFX indicates a special effect, in this case, a window in which each of the guests is seen.

SOT is short for "sound on tape," in this case, music prerecorded on a videotape.

BB is short for "billboard," a full-screen key of a sponsor's product, as you might see at the start of a baseball game or a beauty pageant.

A bumper is an audio or video device that ends a segment, and usually leads into a commercial. They're used regularly on "The Tonight Show" and "Late Night with David Letterman."

Opening Billboards, Host Intros

Video	Audio
Beauty shot	*Anncr:* Brought to you by
Key BB#1: Ivory	the makers of Ivory Soap . . . it's so pure, it floats . . .
Key BB#2: Pampers	And by all-new Pampers . . . with a blue ribbon of dryness . . .
To audience	(Music up)
Find Ken	Now, here are the hosts of "P.M. America" . . . Ken Burke . . .
Find Carol	And Carol Kramer . . .
Ken & Carol meet	(Applause)
Ken & Carol to home base	

(Page 2)

Opening   PMA Show #12   Airdate 7/9

Video	Audio
VT #1: Opening animation	(Music: SOT)
	Anncr: On Tuesday's "P.M. America" . . .
EFX: VP in window (VT #2)	A visit with the vice-president, at home . . .
EFX: Ellen R. in window	Our resident expert on families and relationships, Ellen Rosenberg . . .
EFX: James C. in window	Actor James Coco, on his exciting new television series . . .
EFX: Contest VT (#3), then full	And more about the "P.M. Sweepstakes," where you can win over *three hundred thousand dollars!!!*
VT #4: Opening tag	(Music: SOT, then under)
	Anncr: It's the perfect way to end your day . . . it's "P.M. America!!"

(Page 1)

Welcome

Video	Audio
On Ken	*Ken, Carol:* (Ad-lib hellos)
	Ken: Three hundred thousand dollars! That's an awful lot of money. And today could be the day that somebody wins the "P.M. Sweepstakes," and gets it all.
On Carol	*Carol:* We'll be telling you more about the contest later in the show.
On Ken	*Ken:* And then there's the visit with the vice-president—Carol, do you remember when Jackie Kennedy took America on a tour of the White House?
On Carol	*Carol:* I sure do. And I think you'll find this tour every bit as fascinating.
On Ken	*Ken:* Ellen Rosenberg is with us today.
On Carol	*Carol:* And she'll be on with three families with some very interesting problems . . .
On Ken	*Ken:* James Coco is with us today— he's always a terrific guest.

(Page 3)

Intro Commercial	
Video	**Audio**
On Carol	*Carol:* And we'll be back with our special guest, James Coco, right after this important word . . . from Ivory . . .
Audience	(Music up)
Key bumper #66	(Applause)
"Stay with us"	
To black	
	(Page 4)

and other complexities—and then consider that all of this is being recorded live on tape without stopping). On complicated productions, scripts provide standardized information for all concerned (and, usually, a production assistant is assigned to distributing changes to all of those people, lest a cue be ruined because someone had the old script page). Revised script pages, incidentally, must be clearly labeled as revisions—color pages, from white to yellow to goldenrod to salmon to pink, may be used, but big numbers at the top of the page ("REVISION #7") do the job just as well.

STORYBOARDS

While a script can provide may of the details required during the development process, it is not an inherently visual presentation. Words can tell only so much of the story; ultimately, television is a visual medium and only visuals can really describe the finished presentation. This is especially true in commercial production, where storyboards are commonly used to illustrate every scene and scripted words appear beneath each of the illustrations. Each picture-and-word composite is called a "storyboard frame," or just a "frame."

Storyboards are most commonly used in commercial production, where every second, every visual, is carefully plotted to maximize impact. Other types of productions use storyboards as well. Steven Spielberg storyboarded *Raiders of the Lost Ark* in a comic-strip style, to help his crew visualize the entire production and to understand what was on his mind. Many computer programs are also developed using storyboards. Producers of animation regularly use storyboards as writers use written scripts, to get everything down on paper in the early stages, so changes can be made before production begins. Most productions do not require storyboarding, but an occasional use of storyboarding techniques, however rough, can be helpful to a director in planning a complex series of shots or edits.

Storyboard art is usually rough. There is no need for finished illustrations (unless, of course, the storyboard is to be an important part of a presentation to a sponsor, etc.). Marker drawings render the basic ideas effectively, and a pen can fill in the details. While artists may be employed for storyboarding, a director with some basic drawing skills can make a storyboard that will help to plan shot composition, on-screen effects, and edit plans.

Storyboards, like scripts and treatments, are tools to be used and adapted as necessary. When I produced "The World's Greatest Photography Course" for home video, we knew that there would be a great many graphics and individual photographs, and that each would be keyed to a few words of the script. By placing a storyboard frame on a two-column script page, we were able to keep track of a large number of production elements as they related directly to the script and to the storyboard. Our art director saved the work of redrawing the host over and over again by investing in three rubber stamps, each one made from a drawing of the host—one a close-up, one a medium shot, and one a long shot. Each production has different requirements—and a storyboard can be useful as a resource.

Storyboard
A photoboard shows the key frames of a completed commercial spot, along with the scripted material. A similar format is used for the storyboard, which is referred to throughout the planning stages of most commercials. *(Photoboard courtesy A. H. Robins and Jordan, Case, Taylor, McGrath)*

DEMONSTRATION TAPES AND PILOTS

Back to Hollywood. A concept has made it through the treatment and script stages; here's the next step. The idea looks good on paper, but it cannot be evaluated as a television program unless a sample, or "pilot" episode, or several pilot episodes, are made. A few dozen pilots are made every year—there is even a "season" when all of the pilots are made, be-cause they must be delivered in time for the network people to make decisions for the fall schedule—only a few pilots result in a series. This stage of development is a sorting-out process, bringing projects to life as television productions, and then testing the programs with sample audiences to see which ones have a chance in the ratings. Each season, several hundred projects are developed by each of the networks, and only twenty or thirty are piloted. Of these, only a small number are ever seen on the air.

A few years ago, a pilot was a single program, a sample episode in the series, which tested not only the concept, but the casting, the director, the overall approach. Pilots never appeared on the air—except as early episodes of the new series. The environment today is less structured; pilots and test series take many forms. Some go on the air as a made-for-TV movie. Others are produced as limited-run series, where a few episodes are scheduled, and, if the ratings are good, the series continues. In the past, shows were ordered by the networks in full-season lots of twenty to twenty-five episodes. Today, a network may order five or six episodes, and may not air them immediately. So a pilot may not be just a sample episode—it may be several episodes, it may be movie of the week, it may be a miniseries, or it may be a one-time affair.

Outside the networks—and the pay television companies, who follow similar procedures—pilots are an expensive luxury. Demonstration tapes, which are likely to run less than thirty minutes and are equally likely to be produced on a shoestring, are more common. Demo tapes are made for two reasons, to test ideas and to sell ideas. A demo might be made to see how two performers interact, whether three anchors would be appropriate for the six o'clock news in its expanded format, how well a choreographer trained for stage can handle the television medium, how a new talk show plays on television. Demo tapes are more effective, better selling tools than proposals, and cheaper than full-scale pilots, so they are favored by those who sell ideas to sponsors and to local stations (the syndicator waits to make the investment in a full-scale production until a sufficient number of stations are signed—or else decides to drop the whole idea).

There are no rules to producing demo tapes, except to focus on the reason why the tape is being made and to avoid the temptation to create a full-blown production when one is not required. Generally, a shorter tape is better than a longer one.

Demo tapes are also used by talent, directors, and producers as video résumés. This is an extremely common practice in local television, where movement to progressively larger cities in rapid succession is essential for career growth.

BUDGETS

Selling the concept for a new production is only half the battle. Once the idea is accepted, the inevitable question is sure to follow: "How much is it going to cost me?"

The preparation of budgets is a specialty that should be mastered by every television producer. Budgets must be created with a realistic outlook, a thorough knowledge of all aspects of the market, and an equally comprehensive knowledge of the production process. A mental list of potential problem areas is extremely important, as is an understanding of where money might be saved for profit (when you're working on the outside), reputation (when you're working on the inside), and safety ("pad" from one budget line can sometimes be used to cover overages on another).

The idea of a budget sounds complicated. Detailed, yes, but not complicated. A budget is really nothing more than a list of every item that represents a cost to the production.

Creating a realistic budget is best accomplished by thinking the project through with care, and working with a master list of all possible categories (frequently a form supplied by the production company or television station). Either a calculator and an accounting ledger pad or a computer with a spreadsheet program like "Lotus 1–2–3" will be needed as well.

Weekly Production Schedule

A weekly production schedule is developed by working backward from the delivery date, or the date of the first program in the series. Items with long lead times are begun as early as possible (sets and music, on this schedule). Staffs are phased-in as required, usually with support staff at a minimum in the early stages.

On a series like "P.M. America," the concern is guest bookings, and the resultant scripts (which are written interview questions, in the main). On a tape series, tape dates and edit dates would be plotted in a similar fashion.

"P.M. America" Weekly Production Schedule

Week	Week Beginning	
1	6/22	Start key staff. Open production office. Meet with set designers. Meet with composers. Interview writers.
2	6/29	Finalize facility. Hire set designer. Hire composer.
3	7/6	(NOTE HOLIDAY WEEKEND) Develop booking system. Telephones installed. Work out staff responsibilities. Rough set sketches due.
4	7/13	Start all staff. Assign production responsibilities. Revise set sketches. Rough music due.
5	7/20	Start booking week #1 (8/31). Hire announcer. Format meeting with hosts. Finalize set sketches.
6	7/27	Finish booking week #1 (8/31). First draft scripts due, week #1 (8/31).
7	8/3	Start booking week #2 (9/7). Finalize week #1 scripts (8/31). Visit set shop for approvals. Music record session.
8	8/10	Finish booking week #2 (9/7). Week #2 scripts due (9/7).
9	8/17	Set load-in. Lighting. Studio run-through. Book week #3 (9/14).
10	8/24	Blocking and rehearsals. Week #3 scripts due (9/14). Book week #4 (9/21).
11	8/31	AIR SHOWS #1–#5. Book week #5 (9/28). Week #4 scripts due (9/21).
12 (and after)		Bookings are due show #5 weeks in advance. Scripts are due show #4 weeks in advance.

It's best to start by creating a rough production schedule, by making some realistic assumptions about the amount of time required for scripting, casting, scouting of locations, building of sets, recording of music, shooting, screening, editing, and completing the sound track. The result will be a list, with the principal activities listed next to the number of each week (or day, or month) in the production. There is no science here: the ability to create a realistic production schedule is a matter of experience, of knowing how long each of the steps will take, and where the process may go astray. By preparing a schedule as a first step, basic assumptions are made that will be used throughout the budgeting process. If these assumptions change (e.g., an additional week is required for shooting), then the entire budget may change. The value of using a computer spreadsheet program is evident here: a single change can be entered, and the computer will make all of the adjustments required by the change, usually with a single command.

There are many different forms and formats for production budgets. I have created my own, based on a handful of popular formats.

BUDGET CATEGORIES

The first category in my budgets is "staff." I list all of the necessary staff members: writer, producer, associate producer, production assistant, director, associate director, stage manager, secretary—whoever is needed to produce the project. Each is to be hired for a specific number of weeks, so weekly salaries are multiplied by number of weeks to find the budget numbers. Payroll taxes add about 15 percent to the budget (though some staff members are likely to be hired as independent contractors or as personal corporations, so they

will pay their own taxes). If the production is going to be a long one, money for a benefits package should be budgeted as well (local stations and corporations have different policies here, but it is wise to check on any "in-house charges" that may be charged to the production budget). Similarly, production personnel on loan from outside departments are to be paid, ultimately, from the production budget (once again, it's worth asking about in the early stages).

Fees for writers, directors, and associate directors are fixed by the unions, the Writers Guild and Directors Guild respectively, but only in certain markets (these presently include: network television, big-market local television, and some syndication and home video productions). These fees are minimum requirements; union members with extraordinary talent or reputation can demand overscale payments. In some cases, union contracts include "profit participation"; this is common when a production is likely to be seen over the course of years via different media (e.g., when a popular sitcom is released on home video, many of the creative people will be entitled to profit participation—performers and musicians as well as writers and directors).

One last note: not everyone is paid on a weekly or biweekly basis. Some production personnel, and some performers, prefer to be paid on a "flat rate" basis for services rendered. This is always a matter of negotiation, and frequently runs to the advantage of the one doing the hiring.

The "talent" category should cover all on-camera and off-camera performers (e.g., narrators and announcers). Networks and big-market local productions, and producers of national commercials, must employ performers who are members of either AFTRA (the American Federation of Television and Radio Artists) or SAG (the Screen Actors Guild). Both of these unions have clearly established minimums for the markets listed above and for reuses of recorded material; both are currently developing similar rate structures for pay television, cable television, and home video. As you might expect, only newcomers work for scale rates (except in special situations, when the performers have reasons other than money to make an appearance). Once a performer becomes known, he or she usually hires an agent to negotiate fees and residuals. (You'll find more about this in chapter 10.)

Nonunion situations (e.g., medium and small-sized local TV markets) are negotiable, but there are traditions to guide both parties through the negotiation.

"Music" may involve original compositions, arrangements, and recording sessions (all under the jurisdiction of the American Federation of Musicians [AFM], which has offices in most large and medium-sized cities); the payment of fees to the music producer, composer, publisher, performers, and record company for

Budget

A production budget lists every cost associated with a project, organized for convenience in sections. The distinction between "above-the-line" costs (mostly creative, like staff and talent) and "below-the-line"

costs (facilities, travel, etc.) varies with each production house and situation.

This sample budget, from an actual production, shows the costs associated with one hour-long talk show featuring several key executives from a major

WORKSHEETS PREPARED BY HB 10/28/86   1/4

PRODUCTION BUDGET - CORPORATE TALK SHOW   1 HOUR / 1 DAY SHOOT

	SUMMARY		SUB-TOTALS	TOTALS
	ABOVE-THE-LINE			
	STAFF		11460—	
	TALENT		1380—	
	MUSIC		250—	
	DESIGN		3500—	
	TOTAL			16590—
	BELOW-THE-LINE			
	SET		9250—	
	FACILITIES		9030—	
	TAPE STOCK		2295—	
	POST-PRODUCTION		6600—	
	TRAVEL		3000—	
	OFFICE		600—	
	MISC.		950—	
	BUSINESS EXPENSES		1100—	
	TOTAL		33025—	
	BUDGET SUB TOTAL			49615—
	10% CONTINGENCY			4962—
	GRAND TOTAL			$ 54577—

CORPORATE TALK SHOW   A/o 10/28/86   2/4

	ABOVE-THE-LINE ITEMS	COST	SUB-TOTAL	TOTAL
	STAFF			
	PRODUCER — 3 WKS x $1000	3000—		
	ASSOC. PRODUCER — 3 WKS x $750	2250—		
	DIRECTOR — 2 DAYS x $500/DAY	1000—		
	WRITER — FLAT FEE	2000—		
	PRODUCTION ASSISTANT - 3 WKS x $250	750—		
	RESEARCHER — 2 WKS x $225	450—		
	STAFF SUB-TOTAL		9550—	
	+ 20% TAXES + BENEFITS		1910—	
	STAFF TOTAL			11460—
	TALENT			
	HOST — 2 DAYS x $500	1000—		
	ANNOUNCER (V.O.) — ½ DAY x $300	150—		
	GUEST FEES	0—		
	TALENT SUB-TOTAL		1150—	
	+ 20% TAXES + BENEFITS		230—	
	TALENT TOTAL			1380—
	MUSIC			
	RECORD PURCHASES	50—		
	RIGHTS (ESTIMATED)	200—		
	MUSIC TOTAL			250—
	DESIGN			
	SET DESIGNER/DECORATOR	2500—		
	LOGO	1000—		
	DESIGN TOTAL			3500—
	ABOVE-THE-LINE TOTAL			16950—

the right to use a specific piece of recorded music (these licenses are handled under blanket agreements—annual fees that cover the cost of most, but not necessarily all, compositions—by local television stations, but must be negotiated individually in most other situations); the payment of a flat fee for access to a library of "canned" music, or individual selections contained in it. Every production has different requirements in the music department: some must be made with an original sound track (the most expensive option, common in network and pay TV), while others can be effective with a few pieces of prerecorded music from a custom music library, also known as "canned music" (usually the cheapest). Use of hit songs in original productions, except in local stations for reasons mentioned above, can be an expensive habit—even more expensive than having original music composed and recorded for some productions, because the performers, composers, and other creative per-

---

corporation. It was made for internal use, but the producer was told to create a top-quality production. The budget, therefore, was higher than usual for corporate video.

Note that some budget lines read "0"—this is done to remind the client that these items will be provided to the production at no charge.

Note also that this is a working budget, for internal use only. When budgets are submitted for approval, it is best if they are neatly typed.

CORPORATE TALK SHOW   N/o 10/28/86   3/4

BELOW-THE-LINE ITEMS			
SET			
CONSTRUCTION	5000—		
PURCHASES + RENTALS	3500—		
TRUCKING	250—		
PROPS + MISC.	500—		
SET TOTAL			9250—
FACILITIES			
STUDIO RENTAL (W/ EQUIPMENT PACKAGE,			
EXCEPT AS INDICATED BELOW)			
½ DAY SET + LIGHT, ½ DAY STRIKE	1500—		
1 DAY SHOOT	2500—		
SUB-TOTAL		4000—	4000—
CREW:			
3 CAMERA OPERATORS x $250/DAY	750—		
1 AUDIO ENGINEER x $250/DAY	250—		
1 T.D. x $300/DAY	300—		
1 L.D. x $325/DAY x 2 DAYS	650—		
2 STAGE HANDS x $200/DAY x 2 DAYS	800—		
1 STAGE HAND x $150/DAY x 1 DAY	150—		
1 STAGE MANAGER x $275/DAY	275—		
1 VTR OPERATOR x $225/DAY	225—		
1 VIDEO ENGINEER x $250/DAY	250—		
CREW SUB-TOTAL		3650—	
+ 20% TAXES + BENEFITS		730—	
CREW TOTAL			4380—
ADDITIONAL EQUIPMENT:			
LIGHTING RENTALS	350—		
GELS (PURCHASE)	50—		
EXTRA MONITOR RENTAL	150—		
EXTRA MICROPHONE RENTALS	100—		
SUB-TOTAL		650—	650—
FACILITIES TOTAL			9030—

CORPORATE TALK SHOW N/o 10/28/86   4/4

BTL, continued			
TAPE STOCK			
3 HRS. 1-INCH x $90	270—		
SCREENING DUBS 15 x $100 (INCL. DUP.)	1500—		
AUDIOTAPE (WORKING STOCK)	25—		
MISC VIDEO WORK TAPES	500—		
TAPE TOTAL			2295—
POST-PRODUCTION			
1 DAY ON-LINE EDIT 8 HRS x $350	2800—		
4 HOURS CHYRON 4 HRS x $125	500—		
4 HOURS QUANTEL 4 HRS x $200	800—		
1 DAY PAINT BOX 8 HRS x $250	2000—		
TITLE CAMERA + MISC.	500—		
POST TOTAL			6600—
TRAVEL			
AIRFARES, HOTEL, PER DIEM (ESTIMATED)	2500—		
LOCAL TRAVEL, CAR MILEAGE	500—		
TRAVEL TOTAL			3000—
OFFICE			
OFFICE SPACE RENTAL, PHONES, UTILITIES	0—		
EXTRA TYPEWRITER - 1 MONTH RENTAL	150—		
COPYING	200—		
POSTAGE + MESSENGER	250—		
OFFICE TOTAL			600—
MISCELLANEOUS			
CATERING $10/PERSON x 25 PEOPLE	250—		
PETTY CASH	500—		
MAKEUP/HAIR - 1 PERSON, 1 DAY	200—		
MISC. SUB-TOTAL			950—
BUSINESS EXPENSES			
ACCOUNTING + PAYROLL	500—		
LEGAL (HOST AGREEMENT, ETC.)	250—		
INSURANCE (1% OF BUDGET)	350—		
BUSINESS SUB-TOTAL			1100—

sonnel all negotiate individual contracts. Still, popular tunes may be vital to the format, as in the syndicated series "Solid Gold," and can be important to the feeling of a production, as in the early episodes of "Happy Days," which featured fifties doo-wop music. When the use of popular music is contemplated, it is usually wise to make a few inquiries as to cost before basing a production on a potentially expensive element.

"Rights" is a category that appears in the budgets for many network and syndicated programs, and in the budgets for an increasing number of home video presentations. The rights category pays the up-front fees for the licensing of specific book titles (like *Roots, The Winds of War, North and South*) and the use of a concept that forms the basis for a production ("Dungeons and Dragons," "Peanuts," "Strawberry Shortcake," "Ripley's Believe It or Not"). Norman Lear paid the owners of a British television series called "Till Death Do Us Part" for the right to produce a new American series called "All in the Family." Sporting events carry rights charges as well: NBC paid dearly for the exclusive right to cover the Olympics, and WPIX in New York pays a negotiated fee for the right to cover the Yankees. If the production is likely to be sold to other markets after it has been seen on network TV (e.g., "Shogun" might later be sold into syndication, or Olympic highlights might be released as a home videocassette), the rights holders generally participate in the profits. In some cases, money paid "up front" is considered to be an advance against later profits.

Many projects are developed without paying the rights fee until the production is under way. An "option payment" will bind the agreement between the producer and the rights holder. An option agreement usually involves an amount equal to roughly 10 percent of the overall rights payment (sometimes more, sometimes less, depending on circum-

stances), for the exclusive right to develop and market the idea for a specific period of time (ninety days, for example). Option agreements are usually renewable for one or two additional periods, provided additional option money is paid. Option money is usually paid against the rights fee; that is, the option money is subtracted from the rights payment. There are many variations, every situation is different, and nearly everything is negotiable.

Staff, talent, music, rights, and other so-called "creative" elements are most commonly grouped as "above the line" budget categories—the term is left over from the days of the Hollywood studio system, but it is still used. All other categories—most are related to facilities and crew—are called "below the line" categories. In many cases, the distinction between the two is a blurry one.

Below-the-line budget items usually are related to facilities, crews, and logistics. The cost of scenery, studio rental, editing suites, office space, office equipment, copying machines, postage, telephones, accounting and payroll, insurance, legal fees, and specialty items like security guards, helicopters, audience bleachers, and trained animals all fall into the general below-the-line area. Calculating below-the-line costs can be complicated, because many of the budget items are provided by the station, by the studio, by the advertising agency. Some of these organizations require budgeting monies only for cash outlays (if the station already owns the studio, for example, in-house projects may not be billed for its use), while other organizations attach an "overhead" figure to a budget. It is wise to inquire about the specific items that will be provided by the financing organization, and how they are to be charged to the production budget. If you are unsure, it is best to budget with the assumption that nothing will be free, and to revise accordingly.

The final budget is a combination of three elements: the above-the-line costs, the below-the-line costs, and an additional 5 to 10 percent of either the entire budget or the below-the-line elements, called a "contingency." The extra percentage is a safety net, to be used only in an emergency, when the producer's budget estimate proves inadequate. In order to avoid going over budget, a producer must be sure to include everything in the budget, to leave nothing out. It's important to consider every detail, from coffee for the crew or a limo for a guest celebrity to messenger service, office rent, electricity and heat, stationery, printing of tickets for the audience —every detail costs money. And a collection of small forgotten items can add up, causing an otherwise well-managed project to go over budget.

After the budget is submitted, it is likely to be revised several times before it is approved. With rare exception, revisions are used to cut the total cost of the project.

How does one cut a budget down to size? In several ways, all very carefully executed. The first, and usually best, way, is to cut down on expectations—to cut a set piece that would have been trouble anyway, to shave a few minutes from a home video or corporate training piece that might have been too long, to eliminate a writer or a staff member and simply have everyone do a little more work. The second, and most tricky, maneuver, is to negotiate better prices without dramatically cutting the production itself. Studios, crews, and editing facilities have become accustomed to the over-budget bellyaches of producers (if the producer is well liked, everyone will try to help; if the producer pulls this sort of thing all the time, nobody will be there when the need really exists). The third, and most dangerous, device is the "wholesale cut"—simply reducing the entire budget by 10 percent or more, doing some negotiating, some moving of money from one line to another, and then hoping for the best. A combination of these three schemes can help to trim the budget.

A final budget does not exist in a vacuum; the bottom line is generally compared with the projected dollar return (in home video and syndication), with the overall budget for the operation (in corporate and educational video), with the likelihood of selling commercial spots (local and network television). If the idea for the project can stand up to these marketing and business concerns, and to the internal politics that affect these decisions, then the budget is approved, and production begins.

FOUR

Preproduction

The period between the first production meeting and the first shooting day is usually called "preproduction," "prepro," or "prep." During this period, every detail, from sets and costumes to the colors of paper to be used for script revisions, must be finalized.

HIRING STAFF

Once the budget is complete, and the necessary approvals are in place, the producer must select a staff. In a local station, in many cable systems, and in most educational and corporate situations, a small number of production people who are on permanent staff move from project to project. Very few people are hired from the outside. Network, pay television, and syndication projects are generally staffed with free-lancers who make a living by moving from project to project—whether the project lasts a few days or several months. Production people tend to move around a lot—in local television, they move to progressively larger markets; in network television, they move from project to project as shows are canceled and others are developed to replace them.

A good producer surrounds himself or herself with the strongest possible staff: people who have solid credentials, who have worked on similar projects before, who have previously worked with the producer or with someone the producer knows and respects. Unsolicited résumés arrive daily; few are ever used in the hiring process. Production assistants stand a chance if their résumés happen to arrive on the producer's desk on the very day that the job is to be filled, but more often find first jobs through personal contacts. (More on this later, in chapter 9.)

A good production staff is a cross between a family and a well-organized small company or department. During preproduction, production, and editing, the hours are long, and the pressure can be extraordinary. Because social plans made with outsiders are apt to be canceled at the very last minute, staff members tend to socialize among themselves, or with people working in similar circumstances. If you've ever been involved with an amateur production in community, high school, or college theater, you've probably sensed the kinds of bonds that develop between people. When a producer puts together a staff, he or she must consider not only individual qualifications, but the ability of the group to work together as a team. Individuals must be loyal to one another, trustworthy, dependable, and respectful of both talents and shortcomings, for the good of the production. The word "chemistry" is often used to describe the staff's ability to work closely together.

Finding the right people to fill the slots in a production staff is not always easy, especially if the project is to be done outside the major production centers (Los Angeles, New York, Chicago, Boston, San Francisco, Pittsburgh, and, for news, Washington, D.C.). When the traditional means of finding people are exhausted, producers look to the newspapers, the radio stations, the magazines, the college media departments, the advertising agencies, and the public relations firms for people with parallel experience. News operations are constantly tapping radio and newspapers for good journalists to supplement the steady stream of reporters whose principal training is not in journalism, but in television performance. Music video is now populated by many production people and executives whose backgrounds are not in television, but in radio or records.

Three key members of the production staff are usually the first to be hired: the associate producer, the director, and the writer. The associate producer is almost always someone who has worked with the producer before (either in a similar role or as a production assistant on a few other projects). The director and writer are hired for a variety of reasons: background in similar projects, good sample or demo materials, relationship with the producer or with people whom the producer respects, and reputation in the industry.

The other jobs, from production managers and talent coordinators to production assistants, are usually filled by a combination of people who have been hired by either the producer or associate producer before, or by personal recommendations. Producers are understandably reluctant to hire people with-

out experience. Most production staffs are a bit too small (a reduction in staff size is one of the most common trade-offs for budget approval), so every member of the staff must be able to do more than a single job. And newcomers require training, which takes time away from more important tasks. So why should a producer hire a newcomer? The most common reason is money—a newcomer will work for less than someone with experience. The best reason is far more subjective—the producer's gut instinct that the novice will add something to the production, and is someone worth developing for the future. There are other reasons why newcomers are hired, some very common—he's just getting out of college and happens to be the program director's nephew; the station's license is up for renewal and there are minority problems that can be eased with a new hire; a college professor has been an excellent source of production people in the past, and this latest candidate seems bright enough.

HIRING THE DIRECTOR

While subjective judgment is an important factor in hiring a director, the success of the entire production is ultimately based on his or her ability, and talent. The job of the director is a complicated one, requiring considerable facility in coordinating many simultaneous events, supervising the entire technical operation of the production from the planning stages through the edit, and, of course, providing instruction and guidance to the performers.

A producer must be certain that the director can handle the project, can provide precisely what the producer has in mind. If the project is a special one, for example, one involving classical music or a sport not usually covered on television, directors with an understanding of that area are likely to be hired. Previous experience also counts: a director who has already mastered six-camera live shoots is more likely to get the job than one who has not.

Most television directors are judged by their ability to create a "tight" production. Camera shots should not only be well composed and in focus, but must appear at the absolute right moment. Cues for the music, graphics, and videotape inserts should be executed without sloppiness or timing errors. Performers should appear in control, never confused or looking toward the wrong camera. Technical errors should be nonexistent on a demo tape. A good talk show director should know when to cover the person who's talking and when to cut away for a reaction shot. Cameras should be placed to cover the action without seeing sides of faces, backs of heads, or other cameras. The director is also responsible for the way a production sounds—music and effects should be balanced with spoken words, loud when it should be loud, underplayed when it is in the background.

Directors oftentimes submit demo tapes with clips from their best work. In such cases, a producer is well advised to request complete projects instead, because they reveal a sense of pacing that cannot be seen in a demo. Complete shows are rarely perfect; producers should take note of the kinds of errors and missed judgments, for these subtleties can indicate a director's skill in working with a crew and in orchestrating the overall studio effort. After screening some samples of a director's work and checking a few references, a personal interview is the only reasonable way to make a decision. If the tapes indicate that several directors can do the job, the decision becomes subjective—who will be the easiest to work with, who comes most highly recommended, who is most likely to work well with the performers and the crew. Then there is the matter of money—some budgets restrict the choice of experienced (therefore, in theory at least, higher-priced) talent. In some situations,

expediency plays a part. Simply being available, with adequate demo tape in hand, can weigh more heavily than a wall full of Emmy Awards if the producer is under schedule pressures. With all of this, the "right" director is hired.

Once the director is in place, the staff is complete, and the planning of the production begins.

THE PRODUCTION MEETING

It would be ideal for the entire production staff to start together and work together until the project is complete or until the show goes off the air. In the real world, most staff members have different start dates because of previous commitments and because of the nature of their involvement. There is no need, for example, to start an associate director long before the shoot; similarly, the postproduction supervisor does not usually start until shooting is about to begin. Writers work through the early part of the production schedule and are usually released when shooting begins; they may see the final version weeks, even months, later at a screening. Internal communication is extremely important.

Regardless of start dates, it is a good idea to set up a production meeting at the very beginning of the project, where everyone gets a chance to meet one another, where the entire production is explained in detail, where specific job assignments are made, where potential problems are discussed. It is here that the production process really begins.

A full-scale production meeting should cover the following points:

1. Format
2. Staff responsibilities
3. Production schedule
4. Casting

5. Technical facilities and requirements
6. Logistics
7. The "look" of the production (sets, animations, etc.)
8. Budget limitation
9. Creative problems

For purposes of explanation, we'll create a typical afternoon talk show, one hour long, intended for syndication. It will be produced in New York City, and will air live on the East Coast and in the Midwest, and on tape delay for the West Coast (to account for the difference in time zones). The show will be called "P.M. America."

1. Format

The sample program "P.M. America" will be highly formatted: the basic structure of each program and the amount of time allotted to each segment will be the same on every show in the series. There will be five talk or on-camera demonstration ("demo") segments, each one between 5 and 6 minutes long. There will be two location pieces, each one 6 minutes long. A national news report will play at 20 minutes after the hour for 5 minutes, followed by 1 minute of local news (produced by the local stations); a news roundup will play at 40 minutes after the hour followed by a 30-second local news report. With the standard open, the standard close, bumpers (commercial lead-ins and outs), and the audience ticket plug ("For tickets to 'P.M. America,' write to Box 1111 . . ."), a total of 49 minutes will be produced daily. Add the 1:30 of local news, and the commercials, and the show will run for 58:45 (the remaining time is used locally for station breaks).

Show Rundown

A sample show rundown, for "P.M. America" #12 (the twelfth show in the series). Note that each segment is numbered, and that these numbers correspond to script page numbers. Once again, this would be distributed on a typed sheet.

1. Opening
2. Billboards, host intros
3. Welcome (Ken & Carol)
4. Intro Comm. #1 (Ivory)
5. Comm. #1
6. Interview/demo: James Coco (Ken)
7. Ticket box plug
8. Comm. #2
9. Intro VT (Carol)
10. VT: Vice-president home (Part 1)
11. Outro VT (Carol)
12. Interview: Author (TBA) (Ken)
13. Intro news
14. News block (Charlie) [See separate news rundown]
15. VT: Tease
16. ***Local news break***
17. Comm. #3
18. Intro Ellen Rosenberg (Ken)
19. Interview: Ellen Rosenberg (Ken & Carol)
20. Comm. #4
21. "P.M." sweepstakes (Ken & Carol)
22. Intro new comedy act (Carol)
23. Performance: Robert & Sue
24. Intro Comm. #5 (Pampers)
25. Comm. #5
26. Intro news review (Ken)
28. News review (Charlie)
29. ***Local news break***
30. Intro VT (Carol)
31. VT: Vice-president home (Part 2)
32. Intro Sheila (Carol)
33. Sheila on vice-president home (Carol)
34. Studio audience segment (TBA) (Ken)
35. Comm. #6
36. Contest plug
37. Tease tomorrow's show (Ken & Carol)
38. Closing billboards
39. Credits

2. Staff Responsibilities

At the beginning of the meeting, the producer introduces the executive producer, who owns the production company. The executive producer is a successful television packager; through the years, he has produced several syndicated programs and a number of network specials. The executive producer, busy with several other projects, and working closely with the syndicator in clearing stations and in arranging sponsorship, will not be involved with the day-to-day operations of the show. He is clearly the boss, the man who signs the checks, who hires and fires. He is also the guiding visionary.

The producer is the man in charge of the production. He reports to the executive producer, who demands a high-quality show that will generate high ratings, and a low-cost production that will generate profits. If these two demands are met, the producer will do well. If the show is consistently shaky in either of these two areas, the producer will be replaced by someone who can make it right.

The producer must maintain an overview. Two associate producers handle details: a senior associate producer is reponsible for content, and a production manager (who has the associate producer title in this instance, an increasingly common occurrence) takes care of logistics and most of the below-the-line budgetary and scheduling concerns.

Working for the senior associate producer is a staff of four talent coordinators who book guests on the show and write the interview segments; four field producers who supply one story every other day; seven production assistants (two in the office planning segments and two on location, while three work in the studio: one in the control room, one with the guests, and one on the studio floor); and three secretaries who work for everyone. Several college interns are available to handle errands and small jobs. If more help is needed, more interns can be hired (they don't cost anything because they work for college credit). "No guarantees on office space for interns—we're short three offices right now."

The production manager has a smaller staff —a production coordinator (who is really a production assistant with a few years' experience) and a production accountant. This group deals with the studio, the crew, and the set, making arrangements for props, makeup and wardrobe, coordinating schedules with the editing facility, and supervising the logistics of the live broadcast. They are also responsible for the budget, the payroll, the checkbook, and the little things, like copy machines and typewriters.

The director has one person working for her —an associate director. During the preproduction phase, the AD and the director work as a team, with the AD concentrating on the details, running technical meetings, making logistical arrangements, setting up the timing sheets and tape logs, and working closely with the production manager.

With twenty-four staff members plus interns, and a daily show to produce, there won't be many opportunities for full staff meetings. Most of the work will be done individually or in smaller groups. In this preproduction meeting, everything possible will be covered so that everyone will have the same information. The first topic is schedules.

3. Production Schedule

Every production is made according to a schedule, which is designed to meet a specific airdate or delivery date. A good production schedule paces the activities of the staff and crew, allowing enough time to plan properly, a sufficient number of days or weeks to prepare for the shoot, a reasonable period to screen all of the taped material prior to the edit. Every production has its own requirements, but the sequence of events is quite similar.

On "P.M. America," some of the creative work has already begun, particularly the work that requires long lead time.

The set designer has already submitted several sketches, with rough budgets attached. He must work long in advance of first rehearsal and the shoot days because revisions are common, and because the scenic shop will require at least a month to build the scenery, a week to paint it, and several days to pack it up and ship it to the studio. Once the set is in the studio, it must be lit (again, over several days because the set is a large one), and, most likely, revised after it is seen for the first time on camera.

The music director has a rough list of the pieces that will be required: the opening and closing themes, music for bumpers (in and out of commercials), music for on-air promos. She is scheduled to play some rough tapes next week, in anticipation of another few weeks of revisions and changes in the show's format (which inevitably create the need for additional pieces). Once the compositions are approved, charts will be written, a recording studio will be booked, and musicians will be hired. Time must be allotted for mixing the tracks, and making some changes as well. All of the music on "P.M. America" will be recorded, and all of it must be available for final approvals before the first day of rehearsal.

The opening animation and the show's logo are also in rough form; the first storyboards for the opening will be available in about two weeks. Once the storyboards are approved, a pencil test will be done (on film) to show the animation in motion. With that approved, the animation studio can complete its work in time for the first rehearsal day as well.

The first rehearsal days are set for eight weeks from today; the first live show will air ten weeks from today. The studio is already booked beginning the week before rehearsal (to permit setting and lighting), and two editing suites are on hold for the field pieces. Target dates are set for final decisions on talent. All of the other concerns fall into place: the first week's guests should be firm three weeks beforehand, the first scripts will be due a week later, the first field piece from each of the producers will be due four weeks before the first show ("banking" segments provides a safety net—just in case someone gets sick, or a guest drops out at the last minute, etc.). These details are subject to constant revision, but that first airdate is a constant. All work done during the preproduction period is in anticipation of the first airdate, recognizing the need to "get ahead," because the second airdate happens only twenty-four hours later, and the third follows immediately thereafter. Getting a show off the ground is one thing; keeping the show going after the first show is quite another. Systems are developed; shows get easier as production becomes a routine.

4. Casting

There will be five regular performers on "P.M. America"—two studio hosts, two location reporters, and a newscaster. The syndicator has indicated an interest in the idea of two female hosts, both in their mid-thirties, but the producer is trying to build a case for the traditional one male/one female approach. Several hundred tapes have been submitted, and he has narrowed the field to six candidates. Although he has his own preferences, he plays a few minutes of each tape at the first production meeting, hoping for some impressions. The director and associate producers are opinionated—they have all been involved in these selection processes before. Some of the more experienced PAs offer impressions; it becomes clear that the producer does not care to waste time listening to the newcomers.

One of the location reporters has already

been cast; her contract on a local newsmagazine show had run out, and she'd always been a favorite of the executive producer. The other location job is up for grabs—there are no outstanding tapes, no front-runners. The producer asks everyone in the room to think about reporters, actors, someone who has a definite personality, with a definite style and a definite look. A "Gene Shalit–type" would be ideal. One of the new PAs, who lived in Cleveland before he moved to New York, suggests a former talk show host there. Someone else knows the name, says he's tough to work with, and the producer nixes the idea. Everyone will keep looking; not to worry, tapes are coming in every day.

Newcasters are no problem, there are plenty of local people around who would love this kind of national exposure.

The producer and the associate producer have been working on a list of regular guests, people who would appear once a week on a variety of subjects, including family life, sexuality, education, and money. The usual names come up. The talent coordinator suggests a few authors who have appeared on other shows, people who could handle a weekly spot. She'll get some tapes.

Satisfied with the progress, the producer moves on.

5. Technical Facilities and Requirements

The selection of a studio, and related facilities, can be a complicated process, because so many details are involved.

"P.M. America" will be produced at NBC in New York, in Studio 8H (the "Saturday Night Live" studio). Although the program is being done for syndication, the production manager has made a deal with NBC for the rental of its available studio space (studio rentals are common at most local stations as well). The NBC deal includes not only technical crew (they're members of NABET, a union for broadcast employees), but stagehands (they're members of IATSE, a stage employees union), pages (ushers, to handle the studio audience), and ticket distribution. NBC could not provide office space, (as is common with some large facilities), and could not provide location crews and the necessary editing facilities. Office space has been rented two blocks from NBC, in the Warner Communications building. Several free-lance camera crews have been signed to year-long contracts. Editing will be done at an editing house located a few blocks from NBC, again on a contract basis. The production manager's assistant will handle the scheduling of crews and editing facilities.

Studio 8H was selected for several reasons. Location was the prime reason: it is located in an office building in Rockefeller Center, a prime spot for visitors to New York City (easy to get an audience). Size was also important: there are very few studios in New York that can handle a big show, and finding the perfect studio space is not always easy. Permanent weekday availability is another important point; NBC can use the studio on weekends, provided that the studio is completely reset for the following Monday. One local station was unable to provide proper facilities because its largest studio was needed for evening news, thereby limiting weekday access. Grid height (the distance from the studio floor to the pipes that hold the lights) was another important concern—the set for "P.M. America" is not only large, it is unusually high. The director sees this as a six-camera show; the studio is presently equipped to handle the camera configuration without modification. Every program has its own requirements, and some of these are limited by the available facilities. In this case, NBC was able to provide everything required.

All of the location work will be handled by five different crews. All are equipped with Betacams (see chapter 5). Since the Betacam is a combination camera and recorder, the producer and production manager believe that three-man crews will do the job. The union insists on four-man crews, and negotiations are under way.

On the basis of this contract, the editing facility is currently installing two Betacam editing suites, which combine state-of-the-art computer editing with a complete range of graphics capabilities. Location pieces will be edited onto a ¾-inch cassette, to permit NBC to play the tapes without new equipment or modification.

The technical setup is pretty standard (with the exception of the Betacam situation, which is becoming increasingly popular). The producer and production manager have selected this configuration because it best suits the production, and presumably because competitive bids from other studios and editing houses did not offer more favorable situations. Cost was only one consideration; their confidence in NBC, in the free-lance crews, and in the editing facility was far more significant. Although most productions are completed under the roof of a single facility (frequently a television station), producers and directors should insist upon the best camera operators, the best switchers, the best studio space, the best equipment, the best location crews for their productions. This insistence may sometimes run counter to a company's policy on, for example, the fair distribution of choice projects between all of the station's camera people. It is in gray areas like these that a talented producer, production manager, or director somehow manages to get what he or she wants, while maintaining the friendship of all concerned. Being a producer or director is not simply a matter of handling the studio or the control room—it is handling all of the situations, all of the people who contribute to a project's ultimate success. Whether it's a scheduling coordinator, a shipping clerk, or a secretary, a good producer makes certain that everybody is interested and involved. Helping hands are extended from some very unexpected places, particularly in corporate video, local television, and basic cable productions. It's important to remember that this is a people business, and that everybody likes to feel that they're part of the show.

6. Logistics

Much of the time between the start of preproduction and the first shooting day is devoted to details. The logistical plan is every bit as important as the technical and creative plan for the show; poor logistical planning has been the cause of many difficult shoots, and is frequently the reason why projects go over budget, and why projects are delivered late.

Every project has different logistical requirements. On a six or eleven o'clock news program, for example, camera crews must be carefully coordinated with reporters' schedules, so that crews are not "double-booked" (e.g., the crew is supposed to be on the east side for one story at 3:00 P.M. and on the west side for another story at 3:30 P.M.). On a program with a studio audience, like "P.M. America," plans must be made for holding those people outside the studio until the appointed time—having them stand in line in the parking lot can be a real problem if it rains. Teleconferences require considerable logistical planning, to coordinate studio schedules with the schedules of the executives, and to be certain that all of the operations in different cities are properly linked to either microwaves or satellite systems.

On a production as complicated as "P.M. America," there are a vast number of logistical details. Some have already been solved—there is already a "foolproof" system for moving completed location pieces from the editing facility to NBC, for example. Working at a place like NBC, where so many different kinds of shows have been done, makes life easy. The makeup rooms are large enough to accommodate several guests; the green room (the holding area for guests—it's almost never green, but it's always called a green room) is comfortable, and is already equipped with a monitor (so the guests can watch the show), telephones, and even a steady supply of coffee, tea, Danish, pound cake, and bagels. A member of the production staff will be assigned to greeting guests (and securing signed release forms [see chapter 10] prior to their appearances). At some point prior to the appearance, either one of the associate producers will introduce himself to the guests, officially thank them for coming, and explain how the interview will work. The host will meet most of the guests in the green room, or in the makeup room before the show begins. The audience will be held downstairs, far from the studio, until the appointed moment, when a separate elevator is used to transport large groups to the eighth floor. The staff need not worry about the audience: NBC's pages have been handling this sort of thing for years (even so far as heading down to the street to find extra bodies for the audience). NBC has been provided a few small offices for staff members who need a quiet place to rewrite scripts, book last-minute replacement guests, etc. Without these facilities, one or more staff members would have been assigned to make similar arrangements—supervising the setup of appropriate makeup facilities, ordering coffee for the green room, etc.

On a production as large as "P.M. America," logistics involving the set and wardrobe are handled by specialists. On a smaller production, particularly one in which the budget is just large enough to cover the shooting expenses, these details become the producer's problem. Many new producers get so mired in logistics that they lose the overview; experienced producers try to do as little of this work as possible, finding production assistants, in-

terns, secretaries, or friends of the production to handle the details.

7. The "Look" of the Production

Every production has a "look." Most "looks" are defined by the type of production. News programs, for example, look the same no matter where you go: one wraparound desk in the foreground, with a few flats and either rear-projection screens or chroma-key areas (see chapter 7). Talk shows usually require a few sturdy chairs, a small coffee table, and a platform, with a few plants and either flats or a lit cyc (this is explained in chapter 6). Even home video exercise shows seem to have a standard look: a few men or women in leotards, working out in either a real exercise room or a studio dressed to look like one.

"P.M. America" will look like a homey talk show, in the tradition of "Good Morning America," "Hour Magazine," and similar productions. The set will have a warm look, with wood and plants and a few accessories reminiscent of a country home. The designer is thinking about a country-home-in-the-city feeling, somewhat cosmopolitan but at the same time relaxed. (These kinds of contradictions seem to work on television; they'd never get by in real life.) The hosts will dress in a "designer-casual" style; tweedy sport jackets, everyone in a studied informality except the news people. The show should feel comfortable, inviting, and welcome. All of the music, the show logo, the promotion, the advertising will suggest this easy style.

The producer has every confidence in the designer's ability to create this environment. The director's concerns are aesthetic as well, but she is concerned about the way that the set will affect the cameras' ability to cover the action. "Coziness is fine," she explains, "but we must open the set up, flatten it a bit, if we are going to be able to get close-ups of one guest without having another stick his nose in the picture." Directors must work closely with set designers. (This was never more true than during the disco boom of the late seventies, when mirrors were important to the "look." Mirrors reflect everything in sight, including lights, cameras, and crew. The directors helped the designers determine angles that would permit coverage while maintaining the "look.")

If the production cannot afford a designer, both producer and director are best instructed to avoid complicated arrangements and unusual backgrounds (e.g., reflective surfaces, slender plants that may move with the currents of the studio air conditioning), and to select furniture, flats, and accessories that will wear well, and will not require maintenance, long setup, or touch-up (because of fingerprints, etc.).

8. Budget Limitations

"There's never enough money to do it right, there's only enough to do it again." That's one of many quips made by directors who are limited by too-tight budgets, who may see their "best I could do on a limited budget" efforts redone by another director who has enough time and money to do it right the second time. Limited budgets can be tricky, but their effects can be minimized with clever planning and a realistic view of what can and cannot be done for the available money.

A clever producer-director team can make every dollar go far by hiring people who can do several different jobs, by depending on old tricks, by working with facilities who offer a few extras. A producer or director who creates a difficult working environment because of budget limitations is doing the project a disservice: you work with what you've got, and try to stretch it every way you can. (If you can't work with the budget, don't take the job.)

Once you've accepted the job at a particular budget, it is considered quite unprofessional to ask for more money (it is done, but not often, and only under extraordinary circumstances).

Every production has budget limitations; just about every project could benefit from an extra production assistant, an extra associate producer, a few more hours of camera rehearsal, one more studio day, another day of editing. A good producer fights for the best possible budget, and makes reasonable compromises that will permit the project to begin production under circumstances that are acceptable to all.

A producer working on a series is well advised to work with a staff that's large enough to get the project going. Once the routines are established, the staff can be made a little smaller (fortunately, this is accomplished by attrition: people find other jobs, grow frustrated with the amount of work involved in the start-up, etc.).

Even "P.M. America" is subject to budget limitations. The project was originally conceived as a show with live daily feeds from five different cities: Washington, D.C., San Francisco, Dallas, Chicago, and Los Angeles. Each of these cities was to have a small studio, with one host for local interviews. The decision to originate the show from New York alone was made because the multicity concept was just too expensive. The production was also going to build its own small editing suites, but the up-front investment seemed risky (after all, the show could go off the air after thirteen weeks), so the facilities will be rented instead.

9. Creative Problems

With so many creative issues to be resolved on even the simplest of projects, the producer must allow time to think through every option.

The biggest creative question facing "P.M.

America" right now is the choice of on-air talent. This is obviously an important decision, because the choices will set the style for the program. If the home audience likes the hosts, the show will probably stay on the air. If they don't, the show will probably go off the air.

Conceptually, "P.M. America" isn't breaking much new ground. There are no comedy skits designed to play for the late-afternoon audience (this was discussed and discarded), relatively few performance pieces (there will be no band, so guests who sing will lip-sync to their own recordings). But it will be an extremely well-crafted talk show, with location segments and news inserts. The creativity is not in the format, the producer explains, it is in the execution. The staff is top-notch, the best in the business.

With that brief compliment from the producer, the production meeting is over. The senior associate producer calls a booking meeting; the production manager makes a call about the live feed and decides to run over to NBC to check it out himself; the director is unhappy about the choice of one of the cameramen, and decides to take the producer out to lunch to discuss it.

Eight weeks to the first rehearsal day. Things are going well. But there's a lot of work to do.

DETAILS

Making a television program is not so much hard work as it is detailed work. A good producer does not get too involved in details—he leaves them to the associate producers and production assistants. For them, a clipboard or loose-leaf notebook is standard equipment: all of the relevant phone numbers, schedules, floor plans, catering lists should be accessible in an instant.

Detail people are, almost by nature, makers of lists. Some prefer to arrange lists by days,

carrying the undone items from one day to the next (this was my preference when I was a production manager). Others arrange by topics: the set, the music, the guests, etc. A remarkable number of production people just scratch a few notes onto a sheet of legal paper and stuff it into a file. Somehow, this system works.

On a well-organized production, checklists are standard operating procedure. Since the production office and the studio are sometimes located in separate buildings (or in separate areas in the same building), a checklist of items that must be brought to the studio each day is essential (and it should be used every single day, for the one day when the checklist is forgotten, an important item will be as well). In the studio, one or more PAs should be assigned to the small details such as refilling the host's water pitcher and stowing it under the desk, making sure that the backup microphone is hanging just offstage, testing each of the chase theater-marquee lights on the set, triple-checking the order of the index cards that contain the interview questions, and the order of the cue cards.

An associate producer usually sets up these checklist systems and then depends on the staff to follow through. When a production person says that he or she will take care of something, that something won't be brought up again unless problems arise. This can be tough to understand, especially for people who are accustomed to close supervision. There's just too much to do on a television production for everyone to double-check one another. The word "assume" is one of the most dangerous in the production business, especially when used in a statement like "I assumed he was taking care of that . . ." when an important guest is waiting at the airport without a car to pick him up, or when there are only two cameras in the studio and three are needed ("I assumed that every studio had at least three cameras"). When a prop is missing at showtime because one of the PAs "assumed" it was there, the entire production loses momentum as someone runs around the building searching for a replacement. More often than not, the assuming PA gets a severe reprimand (the first time), or gets fired (the second). "Bear in mind," someone once told me when I assumed something as a production assistant, "the first three letters of the word 'assume.' "

The principal details on "P.M. America" are related to guests. Every show will have five guests. Some will be pitched by book publishers and the public relations community, but most will be suggested by the staff at weekly booking meetings. Some guests are selected because they're well known (the movie stars, television stars, other celebrities), others because they've done something newsworthy (the inventors of a popular board game, the authors of a new book about beauty). The producer will frequently suggest a "type" and the staff will follow with suggestions (e.g., "The Anniversary of the Boy Scouts is coming up, can we find a few celebrities who were Scouts?"). Once the decisions are made, the associate producers "preinterview" the guests (usually by phone), and then write questions for the on-air interview. The date of the appearance is set and the guest is told how to dress, whether to bring any props, and how to find the studio. Guests travel at their own expense, or, more often, at the expense of the book publisher, public relations company, etc. Once a guest is "booked," his or her name appears on a large schedule board ("the booking board"). The associate producer usually calls the night before the appearance to confirm, rather than "assuming" that the guest will show up at the appointed hour, with all of the necessary materials.

With guests, and with other details, it is usually wise to work several weeks ahead of the airdate (or tape date). Two to three weeks ahead is usually comfortable; working too far

Booking Board
A wall-sized booking board helps "Late Night with David Letterman" producers Barry Sand and Robert Morton to visualize future programs. Each guest is represented by an index card, placed in the proper day and in the proper slot on the rundown (e.g., segments 1 to 8, which run from top to bottom on the board). Only guests who are definitely booked appear on the board.

in advance introduces the possibility of a change in the guest's own personal schedule. This time frame also allows a show an exclusive interview with a guest on tour; there is a rivalry among shows in the same city to book a good author, celebrity, or other personality before anybody else does. Most shows will not book a guest who is already scheduled to appear on another show in the same city.

The entire preproduction period is, ultimately, devoted to details—making sure that everything is ready when it needs to be ready. The actual work will vary, but the progressively longer hours, the lists, the constant phone calls and interruptions, the frequent meetings, all these are standard operating procedure for any new television production. Things get easier with time, provided, of course, that the show stays on the air for a while.

When "P.M. America" was still in the planning stages, the senior associate producer was to have a staff of four talent coordinators and five PAs for office and studio work. The work of planning the various segments has turned out to be more time-consuming than planned, and one of the PAs is now doing nothing but talent full-time. At the same time, the PAs found that they could redistribute the work load to allow each of them to handle a balance of studio and office responsibilities on a rotation. This met with the producer's approval, for his staff is more versatile than before.

Once a show is on the air for a while, people tend to figure out how best to handle responsibilities, and should be permitted to do so.

THE PRODUCER PREPARES

During the preproduction period, a producer is very much a manager, careful with his people's time, keeping his eye on the calendar, assuming ultimate responsibility for every arrangement made by his staff, every claim of what will be shown on the screen.

Author: I'll only appear if you show my book. I've been disappointed before. . . .

Associate Producer: I'm the associate producer. If I say that we can show the cover of your book, believe me, we will show it. For ten seconds if you like.

Producer (*sometime later*): We never show books on the air. Call him back and tell him that we'll mention the name, but we can't show the book. It's station policy. And from now on, please check with me before you make that kind of commitment.

Many producers choose to be quite harsh in the early stages, just to break in the staff properly. Occasionally, they must replace the weaker members. When there are two new PAs, both fresh out of college, and one does not pick everything up as quickly as the other, the producer has every right to replace the one who's slow to learn. If an associate producer

agrees to take a job, "but only if I can finish the edit on another project in the evenings," and those evenings run until five and six in the morning, the producer must make it clear that there are responsibilities to this project that must be met, regardless of outside commitments.

A good producer spends a lot of time playing psychiatrist, being boss, jumping in when a booker is out sick and tomorrow's show has a hole in it, or rewriting a segment that's particularly troublesome. Many producers have a particular gift for working with performers, understanding the insecurity, the love/hate relationship with notoriety, the insistence on having everything right ("I'm the one who has to go out there every night!"). Other producers are visually gifted: everything they produce has a distinctive "look." Still others have a knack for timing, or a sense of humor that sets the style for a particular project. In all of these cases, instinct plays an important part, but instinct tempered with the proper training is the producer's greatest asset. The job of the producer is rarely associated with specific responsibilities (aside from keeping the whole operation on time and on budget, and keeping the client or the executive producer or the station happy). Instead, the producer's job is one of manager, part motivational, part organizational, and part creative. Different producers combine these elements to suit their own styles—many of the best producers are not at all creative, but have a knack for hiring creative people (this is very much the case in the movies, for example, and has been for years).

If a producer hires the right staff, he or she really shouldn't have too much to do. If the wrong staff is hired, the producer will be the busiest one in the office, the busiest one on the set, and the one who has to apologize for going over budget. On "P.M. America," the producer has a large, experienced staff. But the careful selection of one or two people on a smaller project can make all the difference.

THE DIRECTOR PREPARES

It has been said that motion pictures and theater are directors' media and that television is a producer's medium. In most cases this is true; the television producer is the one who devises the format, hires the talent, supervises the scripting, makes the important creative decisions. The director may be included in these discussions (more so in Hollywood, less so in local television and most of the other television disciplines). The principal job of the director, however, is to translate the producer's vision into a completed television production. Many successful producers develop projects closely with their directors; as many do not.

Whether the director is involved in the development process or hired sometime later, there comes a point in the production when the director takes the reins from the producer. (I learned this the hard way: as producer, I had been supervising all aspects of the show, and attempted to clear up some points at a final production meeting. The director looked at me angrily and said, "Do your work some other time. This is *my* time now." He was absolutely right.) Experience teaches a producer when to let the director take charge and when to assert himself as the one who is ultimately responsible for every aspect of the television production.

1. The Script and the Shooting Schedule

When a director signs on, he or she should request the current version of the script. If revisions are planned, the director should be briefed. The script is the director's most important tool in the early stages; it contains not only the spoken words and some staging instructions but provides essential information about the project's pace, its look, its technical demands as well. A script in its final stages can also be used to plan a shooting schedule.

Working within the bounds of the production budget, the director estimates the amount of time required to shoot every scene, every segment of the production. Estimates are based on the number of script pages; on the amount of time required to set up, time that might be lost in traveling or arranging for power; on the complexity of the setups and the camera movements (complex setups and camera moves invite errors, so time must be allotted for retakes); on the need for support staff (e.g., makeup, wardrobe—these require on-set time as well); on the likelihood of interference from civilian onlookers; and, of course, on the ability of the performers to handle the material (a new or inexperienced performer might be able to handle only short pieces without making an error).

Taking all of this into consideration, the director arranges the script in a shooting sequence. If most of the shooting is to be done on location, the shooting sequence will be based on the logistics of moving people from place to place (everybody travels to location no. 1, and all location no. 1 shooting is completed there before moving on to location no. 2, etc.). A high-priced performer who has agreed to shoot for five consecutive days would change the priorities; the shooting schedule would be arranged to suit the performer's availability. Fixed times for special events also affect shooting schedules (parades, stage performances, a spectacular sunset). Most location shoots are done completely out of sequence, and later edited into the proper order.

If shooting is to be done in the studio, the shooting sequence may be altered to catch everything on one set before everyone moves into the next set. Variety shows are usually shot out of sequence, to performers' availabilities and the convenience of staging; comedies that are shot before a live audience are usually in sequence, with the stage sets laid out next to one another on the studio floor. Soaps are also shot with many sets in one studio (or two), but no audience, with occasional pieces from location shoots added in the edit.

The schedule for "P.M. America" tends to be the same from day to day, allowing sufficient time for setup, rehearsal, the air show, and any additional tape pieces. Most talk show schedules are not as generous, allowing minimal time for rehearsals (if any at all). The reason: studios must often do "double duty," allowing time for a morning talk show, a commercial, and an evening news program—all in the course of a crew's eight-hour shift.

Shooting schedules should be written to make the most of every available hour of camera time. Travel and setup is best scheduled as the first event of the day; time spent traveling, time spent on logistics, time spent on equipment repairs, time spent looking for a place to eat lunch because nobody checked on it beforehand—these can slow down a production, and eliminate the extra half-hour or hour that would have permitted an important retake. Once the crew arrives at a location, time must be allotted for ESU (electronic setup, or time for the crew to set up the equipment), and for lighting. Once the equipment and the lights are set up, the remainder of the day can be used for shooting (with a lunch break, and with some time at the end of the day's schedule for putting the equipment away).

Marked Director's Script

Although all directors seem to have their own codes and symbols, this marked script is typical.

The handwritten VTR# designations refer to videotape playback machines. "C1" and "C3" refer to cameras 1 and 3. "ESS" is a frame storage device, and the numbers following are individual frames of video, in this case, sponsor billboards. "Sweetening" enhances applause. The numbers in the circles on the left side of the page are shot numbers, the basis of a "shot sheet" (see sample on page 65).

Opening

Video	Audio
VT#1: (Opening) animation *VTR#8*	(Music: SOT) *Anncr:* On Tuesday's "P.M. America" . . .
EFX: (VP) in window (VT #2) *VTR#9*	A visit with the vice-president, at home . . .
EFX: (Ellen R.) in window *C3*	Our resident expert on families and relationships, Ellen Rosenberg . . .
EFX: (James C.) in window *C1*	Actor James Coco, on his exciting new television series . . .
EFX: (Contest VT) (#3), then full *VTR#10*	And more about the "P.M." sweepstakes, where you can win over *three hundred thousand dollars!!!*
VT #4: (Opening) tag *VTR#8*	(Music: SOT, then under) *CUE HIM!!* *Anncr:* It's the perfect way to end your day . . . It's "P.M. America"!!

Opening Billboards, Host Intros

Video	Audio
Beauty shot ⎯C5 Key BB#1: Ivory *ESS #2001*	*Anncr:* Brought to you by the makers of Ivory soap . . . it's so pure, it floats . . .
Key BB#2: Pampers *ESS #2C02*	And by all-new Pampers . . . with a blue ribbon of dryness . . .
To audience C2	(Music up)
	CUE HIM!!
Find Ken C4	Now, here are the hosts of "P.M. America" . . . Ken Burke . . .
Find Carol C1	And Carol Kramer . . .
Ken & Carol meet C6 (HAND HELD)	(Applause) ↖ *SWEETEN!!*
Ken & Carol to home base C2	

Welcome

Video	Audio
C2 ON 2-SHOT On Ken ⟍C1	*Ken, Carol:* (Ad-lib hellos) *Ken:* Three hundred thousand dollars! That's an awful lot of money. And today could be the day that somebody wins the "P.M. Sweepstakes," and gets it all.
On Carol ⟍C3	*Carol:* We'll be telling you more about the contest later in the show.
On Ken ⎯C1	*Ken:* And then there's the visit with the vice-president —Carol, do you remember when Jackie Kennedy took America on a tour of the White House?
On Carol ⎯C3	*Carol:* I sure do. And I think you'll find this tour every bit as fascinating.
On Ken ⎯C1	*Ken:* Ellen Rosenberg is with us today.
On Carol ⎯C3	*Carol:* And she'll be on with three families with some very interesting problems . . .
On Ken ⎯C1	*Ken:* James Coco is with us today—he's always a terrific guest.

2. Working with the Final Script

Once the final production draft is ready, the director can set camera positions and plan shots. Whether the project is a high structured studio show like "P.M. America," a music video, or a commercial, the process is very much the same. The director looks at every page of the script, every sentence, sometimes every word (as in commercials), and decides what the shots will look like, how they will follow one another, and when the camera changes—the cuts, dissolves, and other transitions—will be made. As the director begins to understand the dynamics of the project, how often to cut, the frequency of graphic and videotape cutaways, the sheer amount of time that each performer has on camera, he or she will develop the pace. After the first few pages (or the first few minutes, in an interview show without a script), the pace will begin to repeat itself, and the markings on the director's script will have more to do with cues (when to bring the music in, when to show the graphic) than with cutting cameras.

3. Setting Camera Positions

It is the positions of the cameras that dictate the way that the show looks and plays. Every production has its own unique requirements, but camera placements tend to be quite standard because most sets are, ultimately, either shaped like the letter U, the letter V, or, as in the case of sporting arenas, either a circle or an oval. (See illustrations on pages 60–61.)

Camera positions are worked out on a drawing of the floor plan, preferably an accurate mechanical drawing supplied by the set designer, more often a quick sketch made by the director on a legal pad or on the back of an envelope. Cameras are labeled with the camera closest to stage right as 1, then counterclockwise to camera 3, 4, 5, whatever. Talk and news shows are usually covered with two or three cameras, sporting events can be covered with three or four cameras, but five cameras may be even better. Some televised rock concerts and big sporting events (e.g., the Super Bowl, the World Series) use more than ten cameras.

Cameras should be assigned sparingly—too many cameras are hard to direct, and hard to keep out of each other's way, especially in small studios.

4. Marking the Script

With camera positions set (they will be adjusted in the studio), the director can effectively mark his/her script by assigning shots to each of the cameras and specifying close-ups, medium shots, long shots, one-shots (one person in the picture), two-shots (two people), and beauty shots (a "beautiful" picture of the entire set, the entire event; a wide-angle shot showing everything).

By using all of the available cameras, the director can vary the pictures to create an interesting presentation. If you watch a well-directed television show carefully, you will notice some subtle rules. Cameras are placed so that the viewer never loses orientation. A medium shot of two people talking rarely follows the same shot from the opposite angle—it's disorienting. A sequence usually begins with an establishing shot, usually a long shot, sometimes a medium shot, and then proceeds with close-ups. Sequences are then close-up–close-up–close-up, or close-up–medium shot–close-up.

New directors are well advised to "undirect" a few different kinds of productions. Look carefully at a local news show, for example, and figure out where the cameras are located and why. Draw a floor plan, with camera positions. Notice when the director chooses to cut between cameras, which cameras are used for which positions (on a news show, for example, the movie critic and the sportscaster may use the same desk at different times, and the same camera may cover the meteorologist later in the show), how and when the director uses long shots (usually just before and sometimes just after commercials, frequently with information on the screen, like the day's stock market performance), how many stories are handled by a single newscaster without a graphic, how the shots vary when there's a graphic in the upper-right or upper-left corner. Analyze a talk show ("Donahue" is a particularly complicated one because the host and several cameras are constantly moving), a baseball game ("Monday Night Baseball" is interesting because there are plenty of cameras; a minor league game may be even more interesting because fewer cameras have so much to cover). As you begin to understand how camera positions and cutting work, analyze a situation comedy—one taped before a live audience—to see how the comedy adds the need for reaction shots from other characters, for pacing designed to go for the laugh. In the case of the sitcom, it is important to remember that the show you're seeing is actually an edited compilation of several shows (a dress rehearsal and two plays in front of an audience, plus some pick-up shots, usually close-ups, recorded after the audience has left the studio). "Saturday Night Live" reruns are fun to analyze, because the show is cut as comedy without the benefit of editing, and because there are so many different situations in a single show (news, skits, musical performance, stand-up comedy).

5. Cues

The director's script contains not only camera information; it contains essential information about other cues as well. A "cue" is an instruction to start something, whether it's a videotape, a graphic or sequence of graphics, a piece of recorded or live music, a lighting

Floor Plans

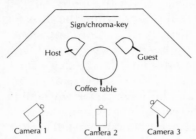

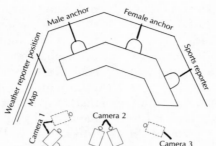

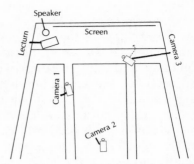

On a straightforward talk show (a), two cameras will do the job, but three are even better. Camera 1 gets a clean close-up and an equally good medium shot of the guest (the long shot will include part of the coffee table and an upright line where the two background flats meet). Camera 2 stays on a cover shot, the two-shot of the guests that appears at the start and the end of the interview and occasionally during the interview to vary the visual presentation. It can be a "locked-off" camera, a camera that's locked into one position by the operator of camera 1 or 3 before the show, and then used without an operator (where the union will permit the use of a piece of equipment without an operator). Camera 3 gets the close-up and medium shot of the host. If there were no camera 2, either camera 1 or camera 3 could get the 2-shot by moving slightly toward the center position. The guests are set up for cross-shooting, so that they may face one another and face the cameras at the same time. This is a standard pattern, used by most talk and news shows.

A news program (b) requires more camera movement. The program begins (and ends) with a beauty shot of the entire set, taken as a wide-angle by camera 1 from the back of the studio (off the floor plan).

Cameras 2 and 3 are likely to be seen in the beauty shot, which enhances the effect of being part of an exciting television studio environment. Camera 2 gets the female anchor, both in close-up and in combination with a corner graphic. Camera 3 gets the same shots for the male anchor. As the show moves on, however, all three cameras move to cover additional shots. Camera 1, at its normal position (solid outline), is used for a wide-angle cover shot just before commercials. Camera 1 also swings upstage (dotted outline) to cover the sports reporter (and any political correspondents, movie critics, and others who sit at the extra desk). Camera 3 gets the weather reporter with the map, while camera 1 swings around to get his or her close-up. While all of this movement is happening, camera 2 arcs slightly to the right, and stays on a wide shot of the anchors as a cover shot, ready to zoom into either one on the director's cue.

All three cameras are equipped with TelePrompTers, so the entire script can be seen simultaneously from any talent position.

The lecture presentation in an auditorium (c) introduces a more difficult situation. Camera 1 plays from a position in the aisle, preferably raised on a platform to meet the eye level of the speaker. Camera 1 will carry most of the

lecture, so it is important that its shot is tight and well composed. Camera 2 gets the speaker and the screen, raised just above the heads in the audience. This camera can also pan left and zoom in for a medium shot or close-up of the speaker. The shots from camera 3, which is best as a hand-held, are added for visual variety. Camera 3 should be able to move on and off the stage, to get some informal side shots of the speaker, wide shots showing both speaker and audience (if their physical proximity creates an interesting shot), and audience reactions. The trick here is to avoid seeing camera 1 in camera 2's shot, and cameras 1 and 2 in camera 3's reverse (audience reaction) shots. If this presentation were to be recorded for later editing, camera 1 would be used as the ISO camera so that a continuous visual could be created for the entire lecture, and the graphics or footage seen on the screen would be recorded before or after the lecture, so the principal reason for the cameras 2 and 3 would be visual variety. In situations where the screen graphics are recorded separately, camera 2 could be eliminated (provided that camera 3 takes a few cover shots into which the screen images would be edited later).

Cameras at a field event like football, soccer, hockey, or basketball (d) are placed along one

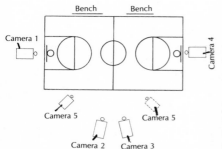

high position, camera 3 spots key players and handles extreme close-ups (you might think of camera 2 as the play-by-play and camera 3 as the color). Camera 1 gets the shot of players running downfield, close-ups during free throws, and, of course, goals. Camera 4 does the same. Camera 5, hand-held along one side of field level, usually follows key players, and may shoot the bench, the fans, and other ancillary action. (Camera 3 may also be used to shoot the bench.)

With only four available cameras, only the home goal would be covered (eliminating either camera 1 or camera 4). With only three

side of the field and at the goals (placing cameras on both sides flips the field and confuses the viewer). Five cameras are more than sufficient for coverage of basketball. Here, camera 2 follows the ball from a high position, and, also from a

available cameras, the other goal camera would be eliminated, and the hand-held (presently marked as camera 5) would take on this responsibility. Keeping the two high cameras (2 and 3) allows the director to cover all of the action and to follow the key players as well; a third camera makes the coverage more exciting and gives viewers a sense of being at the arena.

Similar camera positions are used for all field sports, with (1) keeping at least one camera high and (2) keeping all cameras on one side of the field (plus the goals) important considerations.

change, or a performance. Prior to the start of production, the director must take the time to mark the script for cues, deciding which word or phrase or camera cut or on-stage action will be the cue to start the music, the opening animation, the opening announcements, etc. On most productions, the director is expected not only to cut cameras, but to cue everyone as well. This can be an enormously complicated job (if you have the opportunity, watch a director do the local news—especially in a major market—and you'll get a good sense of how quickly the director must work, of how many different thoughts must run through her head, and of how completely the fate of the production rests on her skillful performance), so all cues must be written onto the script, usually with a brightly colored marking pen so they will not be missed.

6. Working with the Associate Director (AD)

If the production is a complicated one, the director will not be able to handle all of the cam-

era setups and instructions, all of the cues, and all of the technical coordination single-handedly. The AD assumes responsibility for a wide range of activities, from setting schedules to directing rehearsals, depending upon the situation and the director's needs.

During the preproduction phase, the AD does whatever is necessary to prepare the director, technical crew, and facilities support (e.g., telephone communications) for the shoot. Although these responsibilities vary with the necessities of each project, they fall into three key areas: cues, logistics, and technical.

On shows like "P.M. America," when the director marks her script, the AD uses the information to work out his own script, adding cues for music, for graphics, for videotape playbacks, for lighting changes, for performers, for staging moves, as well as notes that will ease the editing process. Cues are carefully numbered (this is particularly important in a live show, where a hundred different music cues, some as short as a few seconds, are not uncommon); the AD calls for "Music number twenty-eight," instead of "Music that accompanies the opening of the curtains."

Common Camera Shots (and Abbreviations)

Although every director has his or her style and shorthand, certain camera shots are common to all television productions. These shots are used more often than any others, and, in most cases, should be available on demand from any camera operator in a matter of seconds.

Extreme close-up (ECU)   A very tight shot, showing part of a face (the eyes, the mouth, the face without hair or neck), or a particularly detailed view of a prop (like the label on a product).

Close-up (CU)   A comfortable head-and-shoulders shot, most commonly used for interview subjects. A close-up of a prop would show some background, but the prop itself would dominate.

Medium close-up (MCU)   A slightly wider view, showing a bit more of the subject, a bit more background.

Medium shot (MS)   Shows most of the subject, but not the whole body. Particularly useful when the subject is holding something, or when someone else is going to walk into the shot. A medium shot is also used on news shows to show the anchor with a graphics window.

Medium long shot (MLS)   Somewhere between a medium shot and a long shot; usually described to the cameraperson by the director to show something specific on camera.

Long shot (LS)   When the lens pulls back, it also widens the field of view. Therefore, a long shot and a wide shot are basically the same thing. This is also called a "beauty shot," because it is used to show the beauty of the entire scene or stage set.

Extreme long shot (XLS)   Very far away. The shot from the Goodyear blimp at the World Series or the Super Bowl is an XLS.

Camera shots are also described by their contents. Directors commonly create new terms for shots that are frequently used in a particular production (a "doughnut shot," for example, is one that's shot through a hole in the set). The terms listed below are common to just about all productions.

Two-shot   A shot showing two people (dogs, cats, products).

Three-shot   A shot showing three people (etc.).

Full body shot   A shot showing someone's whole body.

A good director talks his camerapeople into the best possible shot composition. Adding "headroom" increases empty space at the top of the screen (the cameraperson accomplishes this by "tilting down"—tilting the head of the camera down adds picture to the bottom of the screen). "Tightening" or "pushing in" (also: "zooming in") makes the subject look larger. "Loosening" or "pulling out" (also: "zooming out") makes the subject look smaller.

Everything is coded, everything is coordinated by the AD. The most complicated shows will have several ADs, because there are so many cues to be coordinated.

The AD also sets up the control room and the studio so that the operation will run smoothly. In a logistics meeting with the studio manager, the AD might bring up anything from telephone lines to the placement of monitors in the studio or control room. The AD should think of everything, and, preferably, take care of it during preproduction, when there's plenty of time for details.

Technical arrangements should be checked by the AD as well, working closely with the production manager, making certain that the entire operation is properly coordinated, frequently handling details directly (the AD is actually in the control room during the show, but the production manager is usually planning the next day's shoot or edit). If a character generator is needed for the production, the AD would discuss it with the production manager or the studio manager, leaving the director to more important concerns. If one of the technicians is replacing another who is out ill, the AD would talk through the show, again leaving the director free to solve bigger problems. The AD serves as a communications coordinator as well, making certain that all of the technicians have up-to-date schedules, show routines, and, in the case of the character generator operator, an up-to-date list with properly numbered frames (see chapter 7).

A top AD can make a good director look even better, by planning, by thinking of everything. By "encouraging" the director to storyboard every shot, an AD can avoid a situation where a dozen creative and technical geniuses suggest the best way to have the model move her hand so she does not obscure the Cheez Whiz at a few thousand dollars per studio hour.

If the budget permits, it is preferable to have the AD join the director during the preproduction period. In the real world, however, most ADs start work only a few days (even hours) before the shoot. Experience, therefore, is essential. An AD must be able to walk into a control room an hour before rehearsal, or an hour before air, and organize the entire operation, make certain that the lines of communication are working, that cues are coordinating and there is a system in place for their use during the show (and that they can be heard and understood), that all of the technical people are familiar with the director and her style, that the whole project will move along without logistical or technical problems.

Once the production actually begins, the AD will concentrate on cues, sometimes setting up specific camera shots (e.g., a complicated split screen, where the AD talks each camera operator into position while the director is concentrating on the live action), communicating with the tape operators and with other specialists (e.g., the character generator, the slo-mo disk operator, sound effects). The AD will also concentrate on the clock, using a stopwatch to time specific segments and to count into and out of videotape playbacks, commercials, and other segments originating outside the studio and control room.

In all network and most major-market local television operations, such as "P.M. America," both the AD and the director are hired under DGA contracts. On film productions, the assistant director runs the set and is responsible for the crew; the production manager handles the logistics from the office. On tape productions, the AD assists the director in the control room, on the studio floor, and, sometimes, in the edit.

7. The Crew Meeting

When the director first walks into the studio, he or she should introduce himself/herself to the technical and stage crew, and should immediately organize a crew meeting. In the meeting, the director should talk through the entire production, from the layout of the set to each of the shots to the positions of the microphones, stopping to answer questions as specifically as possible. Even a daily news program should begin its studio day with a crew meeting, even though it may only be a brief discussion of the day's unique requirements or a recap of the problems that should have been solved since yesterday's show. On "P.M. America," the day begins with a meeting to talk through the rundown, and to review any special technical or logistical arrangements.

Once again, it is wise to avoid assumptions: equipment that worked perfectly before lunch is not certain to work perfectly afterward; a slate that is usually left on the monitor may not be there (because the person who usually leaves it has been reassigned to another shift); the host's key light (the primary source of light on the host), which has been in the same position for over a year, is no longer there at all (last night a location crew needed an extra instrument, and nobody was around to ask, so they just took it, planning to return it when they get back to the city late next week). Every show should be treated as the very first show; everything must be double-checked. The rehearsal may last only five minutes, but it should be long enough to check all of the basic studio functions. (There's no need to go through the whole script, though it's usually a good idea if time allows.) The crew meeting should also be used to keep everyone interested, to talk about ways to improve the show periodically. Television production should never become routine—when it does, people get sloppy. Local stations, where only two or three directors work with the same crew people day in and day out, tend to skip crew meetings, and the results often show on the air. Every production should also be preceded by a rehearsal, even if the show has been on the air for twenty-five years. The rehearsal may be brief, as is often the case on local news and talk shows, but time should be allotted to run through a few key elements, just to be sure that everyone is sharp and all of the equipment is working properly.

On most local news shows, there is no time to screen every pretaped segment, and even the director must work from scripted outs (the last few words of the piece are called the "out" or "outro"—the opposite of "intro"). Here, the director must be sure that everyone knows which machine each piece is on. Because there are so many tapes playing in rapid succession (¾-inch playbacks, instant-roll 2-inch machines, 5-second roll 2-inch machines, and a 1-inch machine) as well as a slide chain and a film chain and some special-effects devices, it is vitally important to talk carefully through the entire show, noting each of these inputs and when they will play, with the switcher or technical director (who will make the visual switch from machine to machine), the tape operators, and the audio engineer (who will take the audio feed from the proper machine). This kind of coordination is essential. A good AD's work before the show—making certain that everyone has the up-to-date information that they need—can really make the difference and prevent serious errors. With few exceptions, there is no time during a live show for the director to give detailed instructions to each of these engineers.

On a big one-shot project, the crew meeting may take several hours and may be scheduled in an office, days or even weeks before the studio production. It may be divided into several sections, preceding, for example, each section of the show as it is taped. Crew meetings are tools for the director, to be used as needed. Some crew meetings are run by the AD, who acts, as always, on behalf of the director, as would be the case on later shows in the "P.M. America" series. (Incidentally, on a new show or any complicated project, it's a good idea to prepare a kit for every crew member, a manila envelope that includes show routines, schedules, contact lists with home addresses and phone numbers, even a blank pad to take notes—precisely what he or she needs, and no more, because unnecessary information may confuse rather than inform.)

8. Shot Sheets

On a production where many camera moves and a wide range of shots are required, camerapeople look to the AD for "shot sheets." A shot sheet is, very simply, a list of camera shots. Working with the director's final script, the one with each of the camera shots marked, each of the cuts clearly planned, the AD prepares individual lists of shots, in order, for camera 1, camera 2, and so on, for each of the camerapeople. Shots are numbered (e.g., camera 2, shot #14), in the AD's script and in the director's script. The AD discusses each of the shots with each of the camerapeople, and the director physically walks through each of them during the early part of rehearsal (setting camera positions is usually called "blocking"). Once the shots are set, they can be preset by the AD by calling numbers ("Camera number two—let's see shot one fifty-four") as they come up in the script. If the system is well

Shot Sheet

A shot sheet is a list of specific camera shots, usually prepared by the director or AD during rehearsal. The sheet saves time—rather than saying "Camera one, find Carol in the audience, she's in the right aisle," the AD or director need only say, "One—shot number six."

Most shot sheets are not quite so neat; they're usually handwritten and hand-revised several times before the shooting begins.

Shot Sheet: "P.M. America"
Show #12   Camera 1

2. James C. to window
6. Find Carol in audience (Camera right aisle)
9. On Ken (MS)
15. On Ken through demo
21. Find "great face" in audience, then to wide shot
28. On Carol (out of VP tape)— demo area
33. On Ken for Ellen R. interview
41. On Ken for sweepstakes plug
50. On Ken for lead-in to news (CU)
52. On Carol out of news (CU, then widen to show audience member)
56. On Carol for VT in (VP—Part 2)
58. On Carol for VT out (VP—Part 2)
60. On Ken to intro news review (CU)
67. On Carol for close
71. To audience for credits

coordinated, and the camerapeople are provided with updated shot sheets, with a minimum number of changes, shot sheets can make the job of the director much easier. If the system is poorly coordinated, and some of the numbers are wrong, everyone will be confused, the entire system will crash, and the words "shot sheets" will promote a groan from everyone involved in the disaster for a very long time.

Shot sheets are common on big shows; still, the system can be adapted, and simplified, for use on small, perhaps complicated, productions. Once again, shot sheets are just a tool to make the job of directing a television show easier.

9. Shooting Schedules

After all of the preproduction meetings, the director finally sets the shooting schedule. Working with an eight-hour day (longer, if the producer or production manager has approved overtime), the crew sets up the cameras and other equipment (about an hour, sometimes more, sometimes less, depending on where you're working). FAX (short for "facilities"— the time when cameras and full facilities are available for the director's use) is scheduled for several hours, generally with a fifteen-minute break. Union rules demand LUNCH in the middle of the day (or close to it—crews on location tend to be flexible about the exact LUNCH hour, provided they do not feel that the director or field producer is taking advantage of their flexibility). After LUNCH, FAX is available until about a half-hour before the end of the day. The final half-hour belongs to the crew, as they pack up the studio for the night.

The day's start time is usually determined by the director and associate producer or production manager. A live program will dictate certain start times, and so will the availability of studio audiences (if the audience is to include children, shooting must be done during the day because kids tend to get restless at night). The way that the FAX time is divided is also determined by the producer and director—a certain amount of production must be accomplished by day's end, and the schedule is built to accommodate the goal. Shooting schedules are more than goals; they are precise maps of the day's activities. Because running even fifteen minutes late can cause serious logistical problems as well as overtime, many producers and directors are stick-

Daily Studio Schedule

A sample schedule for "P.M. America." Copies of this schedule are distributed on the preceding night, sometimes with special notations (e.g., the time of individual performer's calls for makeup, the lighting crew call which may be several hours before the start of the day, and so forth).

Daily Studio Schedule
"P.M. AMERICA"

(Effective: Monday, June 30, 1986, Studio 8H)

9:00A—10:30A	ESU
10:30A—12:30P	FAX, Blocking & Rehearsal
12:30P—1:30P	LUNCH
1:30P—2:00P	RESET
2:00P—3:30P	FAX, Rehearsal and Pretapes
3:30P—3:45P	Audience In Crew Break
3:45P—3:55P	Audience Warm-up
4:00P—5:00P	LIVE SHOW
5:00P—5:15P	RECORD PROMOS
5:15P—6:00P	Audience Out, Cleanup
6:00P	WRAP

lers about staying on schedule. Somehow, when time is lost, it is never regained. In fact, a schedule that starts late tends to drag throughout the day. Starting on time, being properly prepared, setting up the schedule with a realistic understanding of what can be done—all of these are important to completing the day's work on time, and to completing the entire project on time and on budget.

10. Crew Calls

Once the studio or location schedule is set, calls are issued. Everybody involved in production has a "call"—a time to be in the studio, or on location, ready to work. A morning call of 6:00 A.M. means arriving at the studio at 5:45 A.M. Calls are sacred; people who arrive late for calls disrupt the entire production, feel very foolish (when fourteen people are all waiting for you before they can start, nobody will mince words), and cause an uncomfortable situation on the set. Returning from lunch on time is equally sacred; nobody arrives late because "the waitress was slow" or "we couldn't find a restaurant." There are no excuses—the show must go on (ON TIME!).

Generally speaking, the lighting crew has the earliest call, because they must be able to hang the instruments and lower the grid, if necessary, without cameras, equipment, and other crew members on the studio floor. Once this initial work has been done, the set can be put into place. Then, the lighting continues and the rest of the crew comes in to start the electronic setup (ESU). (If the set is already in place, the lighting crew usually precedes the technical crew by an hour or more, which permits last-minute adjustments of the instruments, which are usually reached by ladder.)

On "P.M. America," the day begins with an hour and a half ESU, followed by FAX for the rest of the morning, to be used for the blocking (e.g., placement) of cameras and performers, and for rehearsal. Pretapes of more complicated segments are done in the afternoon, before the audience comes in. From 3:30 P.M. until 5:15 P.M., the studio is a public place, and little production work, aside from the actual taping, can be done. There is no such thing as "running late" on a live show's schedule, so the director must constantly watch the clock.

11. Rundowns

The director is responsible for the production of a program or commercial in accordance with the strict parameters of time. If the program is to be broadcast, it must be produced and directed with one eye on the clock—the overall length of the show must be accurate to the second.

The rundown is a list of every element in the production, with estimated timings listed next to each element. The director uses these timings as a guide, correcting the course by speeding the talent through certain segments, stretching through others. It is wise to be flexible: the value of a particularly good take that runs a little long or a little short is usually greater than the value of the precisely timed plan. Adjustments are part of the process; if one segment runs long, perhaps some script can be cut from another to compensate. Segments that run long can also be cut down in the edit; oftentimes, weaknesses in the script do not become evident until they are seen on the screen. Sometimes, a producer or director who is forced to edit a few minutes out of the finished presentation for reasons of time is doing the project enormous good. It is difficult to see this, of course, in a rundown.

Show Rundown

The rundown shown earlier in this chapter is a standard format (see page 45). Estimated segment times and running times are easily added as a fourth column. Actual timings, taken directly from the clock, a stopwatch, or a time code reader (see chapter 5) may be written beside the estimates. A comparison of these figures is essential for the director who must edit a program that must fit into a fixed time slot (in this case, 58 minutes—the program airs live from 4:00 P.M. until 4:58 P.M.).

The very same rundown is used by producers and directors of live productions, where estimated timings must be constantly revised while the show is on the air. This process, known as "back-timing," is explained in chapter 6.

Once again, these sheets are typed. Time codes are either typed or handwritten.

	Segment Time	Running Time
1. Opening	0:30	4:00:00–4:00:30
2. Billboards, host intros	0:25	4:00:30–4:00:55
3. Welcome (Ken & Carol	0:35	4:00:55–4:01:30
4. Intro Comm. #1 (Ivory)	0:10	4:01:30–4:01:40
5. Comm. #1	1:00	4:01:40–4:02:40
6. Interview/demo: James Coco (Ken)	5:00	4:02:40–4:07:40
7. Ticket box plug	0:10	4:07:40–4:07:50
8. Comm. #2	1:30	4:07:50–4:09:20
9. Intro VT (Carol)	0:20	4:09:20–4:09:40
10. VT: Vice-president home (Part 1)	7:12	4:09:40–4:16:52
11. Outro VT (Carol)	0:18	4:16:52–4:17:10

12.	Interview: Author (TBA) (Ken)	4:00	4:17:10–4:21:10
13.	Intro news	0:10	4:21:10–4:21:20
14.	News (Charlie) [news rundown]	5:00	4:21:20–4:26:20
15.	VT: Tease	0:10	4:26:20–4:26:30
16.	***Local news break***	1:00	4:26:30–4:27:30
17.	Comm. #3	1:00	4:27:30–4:28:30
18.	Intro Ellen Rosenberg (Ken)	0:10	4:28:30–4:28:40
19.	Interview: Ellen R. (Ken & Carol)	6:30	4:28:40–4:35:10
20.	Comm. #4	1:00	4:35:10–4:36:10
21.	"P.M." sweepstakes (Ken & Carol)	0:45	4:36:10–4:36:55
22.	Intro new comedy act (Carol)	0:15	4:36:55–4:37:10
23.	Performance: Robert & Sue	3:30	4:37:10–4:40:40
24.	Intro Comm. #5 (Pampers)	0:05	4:40:40–4:40:45
25.	Comm. #5	1:00	4:40:45–4:41:45
26.	Intro news review (Ken)	0:05	4:41:45–4:41:50
28.	News review (Charlie)	1:00	4:41:50–4:42:50
29.	***Local news break***	0:30	4:42:50–4:43:20
30.	Intro VT (Carol)	0:10	4:43:20–4:43:30
31.	VT: Vice-president home (Part 2)	7:45	4:43:30–4:51:15
32.	Intro Sheila (Carol)	0:10	4:51:15–4:51:25
33.	Sheila on VP home (Carol)	1:20	4:51:25–4:52:45
34.	Studio audience (TBA) (Ken)	2:30	4:52:45–4:55:15
35.	Comm. #6	0:30	4:55:15–4:55:45
36.	Contest plug	0:15	4:55:45–4:56:00
37.	Tease tomorrow (Ken & Carol)	0:30	4:56:00–4:56:30
38.	Closing billboards	0:50	4:56:30–4:57:20
39.	Credits	0:40	4:57:20–4:58:00

JUST BEFORE THE PRODUCTION

Every project requires a different amount of preproduction. News programs are prepared daily, two or even three times a day in some markets. Talk shows also must be prepared daily, which means that guests must be booked weeks in advance with subjects of immediate interest written as segments with very short turnaround. Most magazine shows and talk shows are prepared weekly. Network specials may require three to six months or more in preproduction. Most corporate and home video projects are keyed to a delivery date, which usually falls into a target zone of a week or more. Music videos work on extremely short turnarounds; producers submit their script ideas and budgets in one week, and usually shoot and deliver in the span of two more

(this, because the record must be supported while it is still moving up the charts).

After hours/days/weeks of planning, the shoot is set to begin. Everything on the associate producer's checklist has been crossed off; no problems remain to be solved. The director has reviewed the script over and over again, has worked with the performers, has met with the crew. The performers are all comfortable, perhaps suffering from a touch of stage fright (whether this is the second, the tenth, or the thousandth time, most performers get some kind of buzz just before they go on).

"We're starting tomorrow morning at six A.M. And that means five-forty-five A.M.—be there on time, or don't bother coming at all. Take home your phone lists, and keep them with you at all times, just in case there are changes at the last minute. And get a good night's sleep. You'll need it."

FIVE

Field Production

The introduction of portable video equipment in the late 1970s made possible an entirely new kind of television production: "location video." Alternatively called EFP, electronic field production, ENG, electronic news gathering, Minicam, and, most recently, Betacam, location video requires a small crew and a complement of equipment that is easily transported in the back of a station wagon.

With lightweight video gear, a crew with two, three, or four members can move quickly, work in areas previously restricted by physical space or low light, and capture on video a range of subjects that were previously the exclusive domain of film.

THE FIELD PRODUCER

The advent of location video has created an entirely new job category: the field producer. In this position, a field producer fulfills the traditional roles of a television producer, exercising both creative and administrative control, and operates as the director as well, selecting locations, working with the cameraperson to compose and approve shots, super-

vising the technical crew, staging, directing the performers. Once the footage is recorded, the field producer returns to the office, selects the best "takes" and composes an edit plan.

He or she works with the videotape editor and approves all editorial decisions. Most field producers write their own scripts as well: field production demands a singular vision of the completed piece. Unlike television productions created in a studio, location video productions are not collaborative; they are usually the result of an individual's vision. In location video, the field producer enjoys the control traditionally associated with the film director, frequently taking personal responsibility for all aspects of the finished piece from conception and scripting through the edit. This is certainly the case on local news programs across the country; in fact, most reporters are expected to produce their own pieces and appear on camera as well.

THE TOOLS OF LOCATION VIDEO

Since the whole idea of location production is ease of movement from one location to the next, crews tend to travel light, working with a minimum of carefully chosen equipment. A field producer should know something about the basic equipment for several reasons. First, each shoot is preceded by a discussion of the right equipment for the job. Second, there will be times when the equipment and the crew are selected individually (most of the time, the crew and the equipment come as a package). Third, flexibility in the field increases dramatically if the director knows about the latest available equipment. Much of the information about equipment included in this section is very much a part of the studio and editorial processes.

One point of distinction before the descriptions begin. In many situations, the terms "ENG" and "EFP" are used interchangeably. Most of the equipment manufactured for single-camera production in the field is sufficiently lightweight to be used for either fast-moving news gathering (ENG) or productions with less frenzied requirements (EFP). The distinction between the two types of production is only made in situations with very specific requirements.

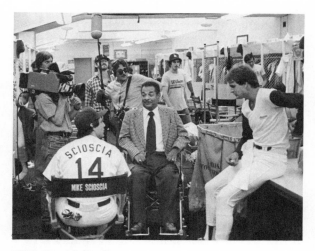

ENG Crew in Action
Barry Rebo and crew cover action in the Los Angeles Dodgers' locker room, as veteran catcher Roy Campanella talks with Dodger star Mike Scoscia. Note the audio recordist behind Campy, with fishpole microphone raised above the interview. A 1-inch VTR is set just to the left of the camera operator (out of frame). This interview was part of a video documentary called "The Boys of Summer," a home video program later released for broadcast syndication. (Photo courtesy VCA Programs, Inc.)

1. Cameras

The basis of all location video work is a one-piece camera that can be "hand-held" (actually balanced on the shoulder) or operated atop a tripod. Ikegami, RCA, and Sony are among a handful of companies who manufacture portable video cameras that are lightweight, balanced for hand-held work, sufficiently durable to withstand travel and some mishandling, and designed for shooting under a wide range of lighting conditions. These cameras are becoming smaller, lighter, and more sensitive every year or so, as designs of the camera pickup tubes are made more sophisticated (a pickup tube is the light-sensitive area behind the lens that translates picture information into electronic information—it is the basis of all television technology). Traditional pickup tubes are being replaced by new solid-state imaging devices, which look like overgrown integrated circuits, and permit dramatic reductions in camera sizes (more than 50 percent). These new devices are more sensitive to every color in the spectrum, sensitive to an even wider number of colors, and not very demanding in terms of power.

Sophisticated camera electronics have also made the job of the camera operator less complicated, as many cameras make the precise technical adjustments—once made by skilled video engineers—automatically. The result is better picture quality, and more time in the field available for shooting.

Camera lenses typically use motorized zooms, with ranges from wide-angle to telephoto, usually at a ratio of 15 to 1. Larger ratios, up to 25 to 1, and smaller ones, down to 10 to 1, are also available and in general use. Although autofocus features are becoming available, most camera operators feel that they can do a better job, a more accurate job, a more selective job (e.g., where to focus, and when) by operating the focus by hand.

Camera viewfinders are small, and usually black-and-white (with small-sized color TV sets for consumers, color viewfinders are sure to become standard in the future). Most crews bring a small color monitor (a monitor is a high-quality television set without tuner and, in many cases, without audio amplifier and speaker) on location shoots, to check color quality and composition on a screen of normal size. The monitor is also used for playback on location, to review a take before moving on to the next location.

For the field producer, the selection of camera equipment will be less important than the selection of the camera operator. Since most operators have their favorite cameras (in larger cities, it is not unusual to find free-lancers who own their own cameras), the choice of camera is rarely a producer's or director's concern. If lighting conditions are to be unusual (e.g., shooting around dawn or dusk, when illumination may be limited), a discussion about the camera's sensitivity is in order. If the shoot will require unusual lenses (close-up, extremely wide or extremely long), this should be discussed as well. Some cameras are relatively heavy and are not easily used on a shoulder for extended periods. Others are lightweight and easy to handle, but there may be other trade-offs. Most camera operators pack a tripod as standard procedure; still, it is wise to double-check, especially if there are special requirements (e.g., the attachment of a teleprompter to the front of the camera). These are all the concerns of the field producer, as well as the camera operator.

2. Videotape Recorders

The second component in the location video package is the videotape recorder. There are three different formats currently in use: ¾-inch (also called "U-Matic," which is Sony's trademark for the format); 1-inch; and ½-inch (also called "Betacam," another Sony trademark, "Recam," which is Panasonic's VHS equivalent, and several other names, all tied to manufacturers' products). The ½-inch format is discussed under "Camera-Recorder Combinations."

A. ¾-Inch Tape
The most common format is ¾-inch; it is widely used for field work on broadcast television and in all nonbroadcast situations because of relative cost and availability. The ¾-inch format uses a cassette: the tape is completely enclosed in a plastic box, cannot become unraveled, and requires no threading. It's easy to store, and it's relatively cheap (about $40 per hour, as compared with $100 per hour for 1-inch and twice that amount for the older, studio-bound 2-inch format).

The downside to ¾-inch tape can be a problem for certain kinds of productions: the smaller-sized ¾-inch tapes (the ones designed for use in the field) are only twenty minutes long. In addition, all ¾-inch images lose quality when dubbed. The twenty-minute problem is overcome by setting up the shoot in chunks that are under twenty minutes long (and changing tapes whenever there's a break in the action or the cassette is almost done), or by using a second machine, which an operator starts just before the other one stops (if all tapes were exactly twenty minutes long, this would work—but because some cassettes are loaded with *roughly* twenty minutes of tape, the operator must start the overlap early, usually at about eighteen minutes, just to be sure.) Dubbing (making duplicate copies of a tape) is a problem on all ¾-inch cassettes, not just those recorded in the field. The first copy (or "first-generation dub") will be slightly inferior to the original; the colors will not be as true and the definition will not be as crisp. A copy of the first-generation dub (called the "second-generation dub") will lose even more color and definition. The third generation will usually show colors that are seriously degraded, and a loss in picture quality that will be noticeable to the average person. Fourth-generation dubs of ¾-inch master tapes are just about worthless; they cannot be broadcasted, and they are rarely even used in demo tapes for office use, so poor is their quality. Technical improvements may rectify the situation—the technology is changing quickly, with ¾-inch tape only about ten years old—but new tape formats such as the ½-inch offer substantial improvements.

B. 1-Inch Tape

Most studio work is recorded on 1-inch videotape. This format is used in the field, but only on big-budget projects because the cost of the recorder, the tape stock, and the editing tends to be several times the cost of the ¾-inch or ½-inch equivalents. It is reel-to-reel format, where the tape must be threaded onto a 1-inch VTR designed specifically for location use (the studio 1-inch machines are quite large and not easily moved; they are found in the field only on the big remote trucks used to cover football games and concerts). The big advantages of 1-inch tape are in the dubbing—fourth- and fifth-generation 1-inch dubs are almost indis-

tinguishable from the original recordings (called "camera masters")—and in the tape loads—up to ninety minutes without changing reels. The additional dubbing capability is important in the editing process, when second-, third-, and fourth-generation dubs are common, especially if the piece involves a lot of special effects (see chapters 7 and 8). The increased capacity for tape loads is important when recording an important interview that should not be stopped (emotional subjects *always* start to cry just before a reel change—ask any field producer), or when recording a concert performance (sporting events aren't as tricky, because there are natural breaks in the action).

If the budget and/or station policy permits, new tape stock should be used for every important production. Every time a piece of tape is run through a recorder (or player), it degrades slightly. After a number of passes, the degradation may become visible on the screen as "glitches" or "dropout" (small white specks that look a little like TV snow). Most stations use tape stock as long as possible; if the production is truly a one-time shot, it is wise to insist upon new tape stock. Bigger budget projects always use new stock because it would be foolish to ruin a $50,000 production to save a few hundred dollars on tape stock. In order to save tape stock in the field (and sometimes in the studio), a director may erase each unacceptable take and save only the acceptable ones. This is a system that should be used with care; the performance in take #10 may be acceptable, but the recording may not.

3. Camera-Recorder Combinations (and ½-Inch Tape)

In the early 1980s, Sony, Panasonic, Ikegami, and several other manufacturers attempted to

Betacam

Music video director Bob Small (in striped shirt) watches for an opening in traffic, ready to cue the driver to start moving. The Betacam rides along, a complete one-man crew that takes care of camera work, video, and audio recording with the press of a button. Since the camera will pan almost 360 degrees on this shot, there is no room in the car for anyone else. Without a Betacam, this shot would be more difficult, and certainly more time-consuming. *(Photo courtesy VCA Programs, Inc.)*

solve the cost versus quality problem with a hybrid package. Technology originally developed for the home may be the answer. Sony adapted its home Beta system to create a hand-held camera and a professional-quality Beta videocassette recorder in a single low-price "camcorder" package. Panasonic's solution is the Recam, a similar product designed with a VHS videocassette recorder. Other systems are based on either the Beta or VHS ½-inch tape formats as well.

There are several benefits to the combination units. First, the entire system (camera and recorder) costs less than half as much as a 1-inch rig that would yield equivalent results. Second, these systems offer picture quality comparable to that recorded on 1-inch videotape, with an equally flexible dubbing situation. The higher speed and the advanced design of the cameras and recorders make this possible. Third, the entire system requires a

minimum of power, and fits into a case not much larger than an overnight suitcase. Fourth, the tape stock is not only cheap ($5 to $10 per cassette, which will run about twenty minutes, compared with the cost of 1-inch tape, at about $100 per hour reel), it is available almost everywhere. Any of the higher-quality home videocassette tapes will do the job (provided, of course, that Beta tapes are used with Beta systems, and VHS tapes are used with VHS systems—the two are not interchangeable). The instant availability of extra tape stock at the local K-Mart has saved many shoots already, and ½-inch professional video is still a young format.

Now the downside. Half-inch recorders are a pleasure in the field, but they require comparable playback and editing equipment back in the studio facility. Many facilities prefer to "bump up" the ½-inch tape to 1-inch tape, just to use the existing editing equipment (a "bump up" is a dub from one tape format to another). As ½-inch video becomes more popular, facilities will buy the necessary equipment. The tape loads on ½-inch systems are only twenty minutes long, which is a problem for reasons discussed earlier. When these two problems are solved, the smaller tape format should become quite popular, especially in local stations, cable outlets, and educational and corporate installations. Right now, however, a variety of formats is likely to be found in most facilities. "Cross-format" editing allows a director to use ½-inch or ¾-inch camera masters, while editing onto a finished 1-inch tape.

Using a ½-inch system in the field is a pleasure. One crew member is both camera operator and videotape recordist, and, with the newer designs, audio engineer as well. Microphones are built into the camera/recorder, or easily attached to it, and volume controls and meters are built into the units as well. For a shoot with minimal lighting and simple audio requirements (e.g., one or two people talking), a "crew" can consist of one person. Although the unions are certain to make their feelings known on the subject (it is to their advantage to keep the crew sizes large, so more of their people work), ½-inch video does permit one person to serve as both director and the entire crew.

One big question remains, though. There are other small-size tape formats on the way, including an 8-mm format whose cassettes are smaller than a deck of cards. These cassettes will permit even smaller recorders and smaller combo systems. The electronics industry is constantly introducing new products, and many of these will find their way into professional video. Some, like the 8-mm design, will come from systems originally planned for consumers. Once again, the information in this book is intended only as a guide. It is wise to keep up with the latest developments in portable video, through engineering contacts (most television facilities and stations have a technical expert who attends the annual NAB and SMPTE shows and will be happy to share the latest information).

4. Audio

One of the greatest advantages of studio production over location production is the control of sound. In a studio, when the stage manager shouts "Quiet, please," everybody stops talking, stops working, stops moving, and the performer is the only one who is heard. The very nature of location production precludes this kind of control: a location shoot should *sound* like a location shoot. Obviously, there must be a balance. At a football game, for example, the crowd should be heard, but should not overtake the play-by-play. A news report at the scene of an accident should be dominated by the words of the reporter, not the screams of ambulances and police cars. A peaceful scene in a suburban backyard setting cannot be dis-

turbed by someone using his lawn mower next door. If you're shooting in someone's home or office, be sure to take the phone off the hook during the interview. Control is the key to good, error-free sound on location.

How is this control achieved? In two ways: by using the proper microphones, and by using your imagination.

A good audio engineer can recommend the proper microphones for any situation. In most cases, the choice will be based on the "pickup pattern" required and just how "directional" the microphone must be. A pickup pattern is the shape of the area that the microphone covers. The most popular mikes are "cardiod"—they pick up in a shape that blends an oval with a heart. Microphones that are very directional pick up only a limited pattern; those that are not, pick up a wide pattern. A directional microphone with a limited pickup pattern would be used by a reporter to cover a noisy accident scene, permitting the viewer to hear the reporter clearly, with the accident in the background. A mike that's not very directional, one with a broad pickup pattern, would be used to cover ambient noise at a football game, or applause in a studio audience.

There are many different styles of microphones available, each with its aesthetic, practical, and technical advantages. These styles fall into three major categories: hand-held microphones, lavaliere microphones (which are small enough to be attached to a tie clip, shirt, blouse, or the lapel of a blazer), and hanging microphones designed to be suspended above the performers, just out of camera range (these are usually the most sensitive because they are used several feet away from the speaker; they're usually the most expensive as well). On location, a hand-held "fish pole" boom offers more flexibility—it is basically a long rod with a hanging mike at the end, handled by an audio engineer like a fishing pole. In many cases, a performer should not use a visible microphone (as in dramatic sequences).

Ingenuity is called for when the guy next door just won't stop his lawn mower, no matter what. Faced with a possible delay of an hour or more, the field producer must consider the options: overtime for a crew of three, the need for an additional shooting day to complete the work (between the travel, the setup, and waiting for the right natural light), or simply paying the guy $20 to "go have a beer." More often than not, the $20 solution is the way to go.

Ingenuity can also be the key to solving unusual audio problems that simple equipment cannot, according to traditional wisdom, cover effectively. A good audio engineer can work magic when teamed up with a field producer who knows the capabilities of the equipment and is willing to take some risks (like sending an expensive microphone up on a helium-filled balloon tied to one hundred feet of Mylar fishing line, which is invisible on camera, to capture the sound of Canadian geese in flight).

While most location shoots require only a single microphone and relatively simple audio equipment, there are several options available for more ambitious shoots. When more than one microphone is used, a "mike mixer" is used to balance the levels of each, to create a single feed to the videotape recorder or for the live broadcast. Wireless (or RF) microphones eliminate cables, giving the performer more flexibility of movement (provided, of course, that the performer stays in the general area of the wireless radio receiver, the other half of the system). There are three pieces to a wireless microphone: the mike itself, which is usually quite small and purposely inconspicuous; the transmitter, which is roughly the size of a deck of cards and is frequently attached by gaffer's tape to a piece of inside clothing, like a shirt under a suit jacket; and the receiver, which is a small box with an antenna located just offstage. The transmitter-receiver connection is made via radio waves. In congested areas like big cities, wireless microphone receivers can pick up extraneous radio signals. When in doubt, it's best to run a test before the day of the shoot.

When covering an event where miking of performers is difficult because they're either too far away or inaccessible (as might be the case with politicians or with individuals speaking from a crowd), highly directional "shotgun" microphones can be useful. The shotgun is held off-camera and simply pointed at the action.

5. Lighting

While there are differing theories about the amount of lighting equipment required for location production, most field producers subscribe to the "less can be more" theory, using natural light whenever possible. Simple productions will usually rent or buy a small lighting kit with two or three lightweight stands and as many lamps, all with barn door attachments (a "barn door" is a set of four large metal plates attached to the front rim of the light that can widen or narrow the beam). The kit fits into a large valise, and is easily transported in the back of a station wagon. The kit also includes various diffusion materials (to soften the lighting), reflective materials (to add some ambience, or to make the lighting appear more direct), and nonreflective materials (to cut down on unwanted reflections). For most interviews and basic coverage, this kit will be all that's needed. If access to electricity is limited (all lights require power), a power generator can be rented as well. If the lighting setup is to be more complicated, as it might be in covering a town meeting or a staged event, additional lights may be rented (if the space is a large one, larger lights will be required, and, most likely, the camera operator will require at least one extra hand to handle the setup). A dimmer board to control the intensity of each of the lights is essential if these intensities are to change during the production (as they do in a stage play).

Lighting on location requires a resourceful mind and a crew that plans ahead so that the lack of a single replacement bulb or fuse does not destroy the plans for an important interview. On one production, where shooting was to be completed long before dark, we found ourselves working at night in a location with plenty of lights, but no access to power. The camera operator solved the problem by convincing the director to change locations so that the light from a store window could provide the necessary illumination. The only problem: the window light illuminated only one side of the host's face. This problem was solved with a collapsible reflector, placed just outside the camera's angle of view.

In fact, the single most important accessory in outdoor shooting is a large white card. If the sun is coming in from the left, the white card can be held just off-camera on the right to bounce the light onto the performer, creating a nice balance. A cue card (20 inches by 30 inches) will do the job. So will a white bed sheet in a pinch.

Generally speaking, overcast days provide the best light. The clouds diffuse the sunlight, eliminating harsh contrasts as well as shadows.

When shooting news at night, the object is to illuminate the action so it can be seen on camera. There is rarely time to set up lights; a single instrument, usually powered by battery and either attached to the camera or hand-held, must do the job of a floodlight. And be careful of car headlights streaking across the background, because the eye is drawn to bright spots on an otherwise dark television screen. More complicated productions require a careful balance of realistic illumination (e.g., from a direction where there is likely to be a light source; a nighttime light comes from streetlights, windows, car headlights, a very limited range of sources), and, usually, a slight

orange or bluish gel for effect. Generally speaking, it is best to work with a minimum of lighting instruments for any nighttime shoot. Too many instruments and too much light from too many sources will destroy the nighttime effect.

Finally, a word about the color of light, or, more accurately, the "color temperature" of light. Outdoor light tends to be slightly bluish, indoor incandescent light, slightly orange, and indoor fluorescent light, slightly greenish. In degrees Kelvin, the standard scale for the measurement of color temperature, outdoor light is about 5,600 degrees, and indoor (incandescent) light is about 3,200 degrees. Since there are several fluorescent light designs currently in use, it is not possible to specify more accurately than the range of 4,500 to 6,500 degrees. Lighting directors and camera operators are well aware of color temperature (which can be measured by a special meter), because it affects the overall tone of the television image. Most cameras are sold with built-in filter wheels that permit the camera operator to filter out any undesirable effects of color temperature. Correction is relatively easy when a location is entirely indoors or entirely outdoors. When an indoor location is partially illuminated by light from an open window, the effects of color temperature can be seen and may be difficult to eliminate.

Outdoor light can be simulated indoors or when shooting at night by using a special kind of lighting instrument that burns at the color temperature of outdoor light. It's called an HMI, and it is relatively expensive, even as a rental item. HMIs are regularly used outdoors as well, to add the color of sunlight to an otherwise cloudy day, and to fill in shadows. The design of these instruments is advanced; one HMI can do the job of several lights of standard design, and can usually be plugged into an ordinary household outlet because it draws less power than instruments of comparable illumination.

6. The Grip Kit

Every television producer and director should be familiar with the single most valuable piece of equipment in any production situation: gaffer's tape. Made of rubberized cloth a few inches wide, this magical material can attach anything to anything. It is used to brace light stands, quiet flapping window shades, dress cables (e.g., secure them so that they're not seen on camera, and so people don't trip over them), insure flawless electrical connections, label tape reels (when someone has forgotten the labels back at the office), hold together pieces of the set when nails, screws, and glues have long since given way. Location crews should never leave home without it.

Besides the indispensable gaffer's tape, the grip kit contains a strange assortment of marker pens, hammers, pliers, string, wire, extension cords, clips, clamps, batteries, and pocket money (for phone calls). Most crews assemble some sort of grip kit based on experience and clever planning; field producers are constantly amazed by the solutions found therein.

7. The Slate (and Tape Logs)

Most grip kits include a slate, which is held in front of the camera to identify each recorded scene and take (as in "scene #12, take #1 . . . take #2 . . . take #3"). A typical slate will include the name of the production, the number of the scene (or script page), the take number (the number of times that the scene has been recorded, beginning with take #1), the record date, and the names of the director, the camera operator, and the production company.

Every take should be preceded by a visual slate (once again, a cue card will do the job if no slate is available), and a voice that identifies both scene and take number (having both an audio and a video slate can save time in editing). All acceptable takes are listed by an AD or PA, and these are later screened and placed in proper order for the edit. With an accurate tape log, showing either each take recorded and its running time, or the time codes where acceptable takes appear (time codes are explained later in this chapter), along with notes ("She turns her head," "The cow moos three times, not twice"), the editing process will run smoothly. Tape logs should be written with as much information as possible in the field while recording, and then filled in later, during less hectic office hours. A copy of each tape log should be packed with each tape, with a second copy in a safe place, and, if many tapes are being used, a third with the edit plan.

8. Continuity

When a location video production is done for later editing, it is important that one person is assigned to "continuity." The term is most easily explained by an example.

The location is a city park, near a carousel with plenty of screaming children. A performer is having a particularly bad time with a particular scene. It's getting late. The director decides that it would be best to start fresh tomorrow. The person assigned to continuity takes careful notes about the scene: the number and arrangement of kids on the carousel, the color of the dress on the lady with the baby carriage, the hand—left or right—in which the host was holding his pipe. Today's footage must be indistinguishable from tomorrow's, and it is the job of the continuity person to make certain that no errors are made. Continuity is usually a PA's function.

Continuity is a holdover from film, which requires time for processing. While proper planning is always encouraged, a continuity person can check details by simply reviewing the videotape just recorded. If the shoot is a

Tape Logs

A. This excerpt from a five-page videotape log shows the AD's notes during shooting. The scenes were recorded out of order.

Scene 12 required three takes. The first one was useless. The second was good, but background noise makes part of it unusable (still, the director told the AD to keep it on "hold" because the reading was good). The third take was perfect.

Scene 16A was a last-minute addition, made as one of the PAs asked about a transition and the director had no good answer.

Scene 3 was particularly troublesome, so the director decided to use the first 35 to 40 seconds of take 3 and the final 30 seconds from take 5, making one acceptable take in the editing room. Take 6 was one final attempt to record the entire 1:10 before the lunch break, but after six takes, the talent was getting a little weary.

Back from lunch, scenes 18 and 20 went very well, but 23 was tricky because the location was noisy. Sneaking takes when the environment quiets down momentarily can be difficult, but the system worked on the seventh take.

The shoot continued throughout this day and the next, and then the director prepared for the edit.

B. With some time to screen, the log is updated to include time codes. The codes were laid down "time-of-day," on a twenty-four-hour clock, so scene 12 began at 10:16 A.M. (plus 10 seconds). Scene 18, which began after the lunch break, began at 1:15 P.M. (plus 25 seconds). In this case, none of the information changed very much with the screening of the buy takes, but the addition of time codes will make the editing session move along more quickly. Note that the AD didn't even bother writing the time code out on scene 23's problem takes; there is no need to fill in every space on a log like this one—only the pertinent information is worth the time and effort.

TAPE LOG

TITLE _PROMO TAPE_   DATE _9/24/86_

CLIENT _LOGICAL EXTENSION, LTD._   PAGE _1_ OF _5_

SLATE INFO _INTERIORS_

SCENE	TAKE	TIME	REEL	HOLD	BUY	REMARKS
12	1	:30	2			Reading error (STOPPED)
	2	:45	2	✓		Bkgd noise (door slam)
	3	:43	2		✓	PERFECT!!
16A	1	:08	2		✓	Transition: enter room (new scene)
3	1	1:15	2	✓		Reading okay, not great
	2	:10	2			Bkgd noise
	3	:45	2	✓		Reading error, good through :40
	4	:15	2			Tech. problems
	5	1:10	2	✓		End is okay, match w/ take 3
	6	:30	2			Talent getting tired / LUNCH!
18	1	:20	3	✓		Good (not fabulous)
	2	:20	3		✓	PERFECT!
19	1	1:40	3		✓	PERFECT!
20	1	:10	3		✓	PERFECT!
23	1	:05	3			Stopped, background noise
	2	:05	3			Bkgd noise
	3	:05	3			Bkgd noise
	4	:05	3			Bkgd noise
	5	:15	3			Reading error
	6	:17	3			Reading error
	7	:20	3		✓	Close enough!

A

TAPE LOG

TITLE _PROMO TAPE_   DATE _9/24/86_

CLIENT _LOGICAL EXTENSION, LTD._   PAGE _1_ OF _5_

SLATE INFO _INTERIORS_   REEL# _2/3_

SCENE	TAKE	CODE IN	CODE OUT	TIME	HOLD	BUY	REMARKS
12	1	10:16:10	10:16:40	:30			Reading error (STOPPED)
	2	10:18:15	10:19:00	:45	✓		Bkgd noise (door slam)
	3	10:20:17	10:21:00	:43		✓	PERFECT!
16A	1	10:41:33	10:41:43	:10		✓	Transition: enter room (new scene)
3	1	10:59:30	11:00:45	1:15	✓		Reading is not perfect.
	2	11:02:05	11:02:15	:10			Background noise
	3	11:04:08	11:04:53	:45	✓		Mistake (good to about :40)
	4	11:07:27	11:07:42	:15			Tech problem
	5	11:41:00	11:42:10	1:10			End of reading okay, match T-3
	6	11:50:00	11:50:30	:30			Talent getting tired / LUNCH!
							REEL #3
18	1	13:15:25	13:15:45	:20	✓		Good (but not fabulous)
	2	13:16:10	13:16:30	:20		✓	PERFECT!
19	1	13:49:10	13:51:30	1:40		✓	PERFECT!
20	1	14:09:15	14:09:25	:10		✓	PERFECT!
23	1	14:21:16	14:21:21	:05			Background noise
	2	14:24:10	14:24:15	:05			Background noise
	3	14:26:30	—	—			Background noise
	4	14:27:15	—	—			Reading error
	5	14:45:40	14:45:48	:15			Background noise
	6	14:51:00	14:51:17	:17			Reading error
	7	14:53:30	14:53:50	:30		✓	Good enough!

B

relatively simple one, with a small staff, this may be easier than taking notes. On a more complicated production, however, time can be saved by taking notes and working with an instant camera; reviewing specific scenes on tape will tie up facilities that could be used to greater value, and may incur unnecessary overtime costs as well.

9. The Location Survey

The survey (also called the scout or scout trip) is a walk through the locations to be used, usually with the owner or manager of the location. Like continuity, the survey is not a tangible instrument, but it is an important tool, used by field producers and by engineers to plan the shoot (attendance by a key technical person is usually a good idea if the shoot is complicated; if the location is fairly typical, like a suburban home or someone's office, the field producer can simply create a checklist with the help of the technical person and report back later).

It is important to explain the logistics of the shoot to the owner/manager of the location of the shoot, providing a clear idea of just how many people and how much equipment will be involved, and for how long, and an explanation of how the space will be used. It is equally important to arrange for all necessary clearances (and payments, if necessary) in advance. The survey should be used to select specific shooting locations for every camera setup, with attention paid to the proximity of power (if no power is available, a gas-driven generator can be carried to the location by van or station wagon), ease of access (to bring the equipment as close to the location as possible, preferably by van or car, or, if necessary, by hand), and general security (making certain that the equipment will be neither stolen nor damaged by onlookers, finding a safe and quiet area for the performers, especially if they're famous). Try to confirm all arrange-ments in writing, and be sure to confirm everything on the day before the shoot. If the survey was a successful one, the crew will simply arrive at the location at an appointed time and start setting up.

SHOOTING LIVE

News programs regularly utilize their portable ENG equipment for live coverage of everything from traffic accidents to press conferences. This live feed is usually accomplished in one of two ways: by hard-wire or by microwave link.

If the location is near the studio facility, a cable can connect the remote unit to the control room. The telephone company can provide hard lines ("telco lines") between any two locations for a fee as well (this requires advance notice).

More common in local news operations is the use of microwave. The remote unit's van is equipped with a transmitter and with a rooftop antenna (it's usually a gold color, and is frequently called a "goldenrod"), or with a dish antenna. In either case, they have a clear line of sight to the station's microwave receiver, free of obstructions from trees, buildings, mountains, and other solid objects. Heavy weather can obstruct microwave signals as well.

Live coverage of an event with a single camera can be a real challenge because the material cannot be edited. Single-camera coverage is usually reserved for a reporter on the scene of a traffic jam caused by heavy snowfall, the background for a weather forecast, or an interview. If the event warrants more coverage, a second van can be sent to the same location, and the director cuts between the two feeds as he or she would do between two cameras. This permits, for example, the reporter to do her stand-up and interviews (remote no. 1) while the viewers can see the cars being hoisted out of the river one by one (remote no. 2). If additional coverage is required, more units can be dispatched (provided, of course, that the station owns more than two vans—this is unusual, even in the largest markets). A director, either on location in a van or in the studio, switches between the two feeds.

SHOOTING ON VIDEOTAPE

Most location productions are shot on videotape for later editing. It is important, therefore, that the field producer shoot every scene with editing in mind. Interview footage should be recorded from one point of view, varying the shot (close-up, medium shot, long shot) from time to time, so that the interview can later be cut into a new sequence without matching problems (it's hard to cut one close-up to another without a "jump cut," which shows the subject suddenly jumping from one position to another). Most camera operators begin each interview take or each new question with a medium or long shot, and then slowly move in for the close-up. Careful attention must be paid to the subject matter. When the subject is talking about something very personal, something very emotional, a close-up is usually best; when the subject is animated and funny, a medium shot showing the hands and some body movement is better than a close-up. Extreme close-ups (tears in the eyes, very shaky hands) can have too much impact and disgust the audience; these are to be used with the greatest of care, and even then in rare instances. Most field producers and camera operators develop a "sixth sense," and simply "know" when to be tight and when to back off. A good field producer will offer some instructions to the camera operator; camera operators who do interviews regularly will not need very much guidance.

Be sure to listen to the interview, and to think about assembling the pieces as they fall into place. Don't think twice about asking the interviewer to rephrase the question (unless, of course, the interview subject is short on time or temper). Sometimes, the interviewee can be asked to rephrase as well. These are important tips—if the interview is not produced well on location, it will be difficult to edit effectively.

If the location piece is not an interview, but a performance piece instead, the field producer has more control. Every take can be perfect. If a performer flubs a line, or there's too much background noise, or the movement isn't exactly right, take it again. And don't be afraid to edit together the beginning of one take and the end of another. Even the most experienced performers require several takes; a producer or director who "thinks editing" can save a lot of time and eliminate a lot of strain for the performers by planning the edit before the shoot. In most cases, the performance will be judged on the finished result. It doesn't much matter how many takes, or how many pieces of those takes, were used to complete the sequence.

REVERSES AND "B ROLL"

When shooting with one camera, things can get pretty monotonous. Visual variety is achieved by shooting supplementary material before or after the interview or performance, and then cutting it in later. Time permitting, one should take some shots of street scenes, crowds, signs (they really help to give a sense of place), anything related to the story that might be used to cover a bad transition from one shot to another. Also include reaction shots from people in the audience, a cutaway to the "scene of the crime," and so forth—all of these can be shown while the interview continues uninterrupted.

The "reverse" shot insinuates a second camera during, for example, an interview. After the basic interview is covered by a single camera, the camera is physically moved to a reverse angle to shoot the interviewer for reactions and a few "staged" questions. When the piece is edited, the audience is tricked into believing that there were two cameras, one on the interviewer and one on the guest. (Some productions have taken this a step further, by recording an interview on location with a producer or writer in place of the celebrity interviewer, who records his or her questions later on. If it's carefully planned and the lighting looks right, it works every time!)

The term "B roll" comes from film editing, where the "A roll" is typically the main story line and the "B roll" contains the background footage. In the days before videotape editing, a news story would be played from two separate film projectors ("A" and "B"), and viewers would see one story line as the director cut live between them. On a typical news story, the B roll contains the background footage (the remains from the fire, the neighborhood affected by the nuclear waste, the cracks in the bridge surface that caused the accident), which is edited into the reporter's running commentary and interviews. Nobody actually records a separate A roll (interviews) and B roll (background). Instead, before or after an interview, the field producer will ask the crew to shoot "some B roll material." The term has changed its meaning from a physical reel of film or tape to a type of footage that should be recorded during almost every location production.

TIME CODES

When a field producer prepares to edit, he or she needs a means of expressing the start point and the end point of the material to be used in the edited master. This is most easily accomplished by using a time code, which identifies every frame on the videotape by number (there are thirty frames per second). A time code reading 01:24:38:12 indicates a position of 1 hour, 24 minutes, 38 seconds and 12 frames.

Time codes can be a little confusing because they can be used in two different ways. The videotape operator (also called the tape op) can start running a code at 00:00:00:00 and record straight through a sixty-minute reel to, approximately, 01:00:00:00 (all sixty-minute reels are a minute or two longer than an hour; few are sixty minutes to the second). If the videotape machine is stopped, the time code can start again in sequence (perhaps skipping a few frame numbers or even a few seconds), or can start again at another time entirely. If there are several stops and starts, it may be best to erase the time code track (time codes are recorded on the equivalent of a third audio track, as digital information), and lay a new time code track, starting at zero, after the tape is recorded.

Time codes can also be laid down as "time of day," which may be helpful in news or documentary production. The code simply shows the time that the material was recorded (this can become confusing if Tuesday's tape looks like Wednesday's, but many videotape operators have developed some tricks that can be used to eliminate the confusion).

On ¾-inch tape, time code is usually recorded on the second of two audio tracks (it may be recorded on an additional "control track" instead). On 1-inch tape, time codes are recorded on a channel designed for that purpose, leaving two audio channels free for sound recordings. Time codes may be seen either on an accessory "time code reader," which connects to the videotape recorder (and is built into larger 1-inch machines) or directly on the television screen. The latter is accomplished by a process called "visual time-coding" or "viz-coding" for short. The code appears in a window most often seen at the top or bottom of the screen and is made a permanent part of the picture. Viz-coded cassettes (also called "window dubs" because of the permanent time code window) are used only for screening purposes; they are dubs of tapes that will later be used to actually assemble the show.

The field producer uses the time codes to indicate the sequence of recorded material to be assembled into the edited master. The codes are typically integrated into an "edit plan," which is both a final script and a list of time code "ins" and "outs" for the editor to follow in his or her assembly. The very same time codes are read by the computerized editing equipment automatically, so the editor may call up any sequence by typing the time code into his or her computer. The use of time codes is the key to computerized videotape editing, as you'll see in the postproduction chapter.

STAYING ORGANIZED

There's nothing worse than returning from a week-long shoot with twenty tapes and discovering that the tapes are mislabeled or the tape logs are missing. The final step in any location production is the double-check. There should be a date, a reel number, and a description on every tape box and on every tape. There should be a videotape log, complete with either time codes, clock timings, or very accurate notes on take numbers, with copies in the boxes and copies on file. The tapes must be kept together (in a library, not under someone's desk). Tapes should be shelved together as well. Projects should be numbered, so that the "SF Giants preseason footage" doesn't get mixed up from year to year, and the president's press conferences are arranged in some kind of order. Oftentimes a tape must be retrieved from the library in a matter of minutes, particularly on local news: NEVER depend exclusively upon the tape library's system—the tape librarian you encounter is likely to be a newcomer (it's a low-paying job) who is unfamiliar with the system. ALWAYS keep track of the library's tape numbers and the titles of each reel or cassette in an up-to-date file that can be easily reached. Control over recorded material is vital; one person on each production should be assigned to monitoring tapes. When a tape disappears, the footage is gone forever and cannot be used for revisions or for future productions.

Although the work can be dull and time-consuming, the field producer should always take the time to review the tape labeling system and the logs, just to be sure that there are no surprises in the middle of an editing session ("Where's reel number twelve B? Wasn't that the tape we relabeled when we realized that we had two reel number elevens?"). In preparing for an edit where a limited amount of material is involved, be sure to take careful notes and transcribe sections if you have the patience (the time "wasted" will always be returned in a savings of editing time).

Dorothy Curley's Tips for Field Producers

Currently a free-lance field producer and writer, Dorothy Curley created many stories for Boston's "P.M. Magazine" that were later seen nationally. These tips for field production are adapted from her instructions for new producers on the program.

Preproduction

Start by thinking through the piece as a story that must be told in an interesting way. Think about the storyteller first.

1. *Storytelling in the first person:* the subject tells his or her own story, usually in an interview.
2. *Storytelling by experts or authority figures:* the official scoop, useful in talking about public figures (and criminals).
3. *Storytelling by outside observers:* usually a subjective rendition, by neighbors, business associates, friends.
4. *Storytelling by omniscient observer:* the on-camera or voice-over host tells the story, seems to know all.
5. *Pictures tell the story:* the director tells the story by deciding what to include and what not to include; given the right story, this is a subjective technique that can be more effective than the spoken word.

A good video piece will use more than one of these techniques to tell the story. Unless you're working with a highly stylized piece, several techniques will be necessary to sustain the viewer's interest and in the case of news to present a balanced presentation.

Now determine the elements—the building blocks. Action is an important building block—there's nothing more boring than long sequences of talking heads. If there is no action inherent in the story, invent some. Move the interview outdoors. Make your subject walk down the street instead of just sitting still. Shoot the subject at home, at work, in the car, with the family, in a variety of locations to keep the viewer interested. Consider stills that can help tell the story and vary the visual content. Press clippings can be used as cutaways or as parts of a montage. Old photographs have a wonderful feel, even if you only use them to introduce a person or relationship. If you find yourself with only one or two locations and no "B roll," think the piece through again. Talk to a few associates and get some fresh ideas. And don't forget about transitions: a good story will move from place to place, and well-planned transitions will help the viewer understand the story's progress. Sometimes, it's a good idea to write the story in paragraph form as you envision the finished piece and to read it over several times in search of better storytelling techniques and opportunities for visual treatment.

On the day before the shoot, confirm time and place with everybody involved—the crew, the interview subjects, the owner of the location where you intend to shoot (and don't assume that the owner has been told everything just because an employee has agreed to be interviewed—this is a common trap that can waste shooting time). Lay out what you want by phone: how much of their time will be required, how long it will take to set up, any special requests (don't ever be afraid to ask for a favor; most people are flattered by all the attention and will agree to almost anything if they feel it can make them look good).

On Location

Before you leave the facility, run through a checklist: blank tapes, the proper microphones, petty cash, release forms, a change of clothes for the performers if the weather looks chancy.

On the way to the location, have everyone travel together in the van and run through the entire story and the production plan and schedule with the crew and talent. Include the length of the story, the style ("plenty of long, lavish shots to show how rich these people are"), the pace, and the plans for the opening and closing.

Once the crew is ready, start looking for details that will tell the story. Signs on buildings, signs on walls, signs that tell you the name and population of the town, people who seem to be indigenous, unusual sights that are routine in these parts, even the interview subject's hands as he or she speaks. Shoot everything with some movement, beginning or ending with a stationary shot. Start wide, then zoom in and stop. Start tight and hold the shot, then zoom out. Pan the camera and then stop. By shooting this way, you can edit either the move or the stationary shot, or both. And get reaction shots, especially if there is an audience, even if you have to stage them by asking everybody to look as though they're listening carefully, or to laugh on your cue. Give yourself options in the editing room by shooting plenty of this kind of material.

Look for tight shots to facilitate editing. Be specific. Use details integral to the story, but avoid the temptation to go for the abstract. Keep it simple.

Be sure you have a good mix of wide shots, medium shots, and close-ups. If the event can be shot more than once, move the camera to a new position for a second point of view. If you're shooting an event, look for more than just the action. Shoot the preparation, the waiting, the warm-ups, other photographers shooting your subject, and *always* get the opening of the event and the close (you can always fake the middle).

During the interview, ask the subject to answer in complete sentences by giving you some of the question back in their answers. For example, if you ask, "What's your name?," the subject should answer, "My name is Jane Jones." Complete sentences and complete thoughts make editing easier. If your subject is on the move during the interview, be sure to catch entrances and exits—stage them if you must—because these will be important for transitions.

On the Way to the Edit

Good edit notes can cut the length of time spent editing in *half*. There is no excuse for wasting edit time by not being *totally* prepared with time codes, and log notes. These notes should include all audio on the tape, verbatim if it will help. If you've never transcribed audio verbatim, try it. The time it seems to waste is always made up later when you're putting the piece together. The notes should also detail the kind of shot, whether the shots have a move (pan, zoom in, zoom out), and how long these shots and moves last. The value of these detailed notes is evident over and over again, like when you've got a 12-second sound bite and you need a shot that can cover.

The edit plan should be written—preferably typed— on several sheets of paper that are handed to the editor along with the camera masters. Every story must have a beginning, a middle, and an end. An edit plan will show you where you're weak. It's a graphic representation of the finished piece. Make your mistakes on paper. It's cheaper.

During the Edit

Prepare your audio track first. It will *always* set the pace for the piece. Most of the audio will come with pictures, so the spine will be in place. Watch the piece straight through. You'll get a sense of timing. If it plays long, fix it before you start inserting pictures. If it plays short, add material before working with additional pictures. Interview sound bites are likely to be the basis for your piece. Never let a nonprofessional voice talk for more than twenty-five seconds without breaking it up with some other sound bites, unless they're crying (crying rivets attention). Use natural sound to stagger the pacing, or use it as background effects to add a sense of place. Use voice-overs by your talent sparingly, to break up the monotony of an untrained voice, to ask some of the more poignant questions, to rephrase something said by the interview subject that didn't come out clearly. Use music and sound effects throughout, to introduce the piece, to punctuate a thought or sequence, to complete the piece. Use music under sequences that establish location. And in a piece told only by pictures, use music to set the pace (once again, the sound track sets the pace, so it must be laid down on the master tape before any pictures are edited).

As for montages, use only tight shots because wide shots will be completely lost. Your music dictates the visuals—a five-second music break demands a five-second video sequence.

Finally, a few rules to keep you out of trouble. Never cut from one wide shot to another. Never cut through a series of close-ups without adding an occasional wide shot as a reference. Never let new characters appear without introducing them first (preferably in some sort of establishing shot, but a lower-third Chyron will do). And never assemble or insert pictures before you have laid down the audio track, and then the basic picture sequence. Build from the basic audio to the basic video to the more complicated video and then finish your audio track. This is the best way to edit any television production.

SIX

Studio Production

There is something undeniably special about the television studio. Even the smallest studio, in a tiny UHF station, holds the promise of creating something wonderful. With the right performers, the right group of people, a studio production, particularly a live studio production, can be exhilarating.

The average television studio is a sound-proofed room measuring about thirty to fifty feet square, with a ceiling height of twenty feet or more. The floor is usually smooth, for easy movement of cameras, and is generally painted battleship gray, which shows very little dirt. The floor may be painted white, black, or any other color as required by the production's set designer, but it is usually repainted gray when the production is over. The studio walls are usually covered with large acoustic tiles at least halfway up; above the panels is raw cement. There is usually one large "barn door" to the studio that is used to load-in scenery and for audience entry (in some facilities, it's called an "elephant door," which goes back to P. T. Barnum's need for a door large enough for Jumbo to enter), and one or two smaller sound-locked doors for access to the control room and the rest of the facility. The studio ceiling is twenty feet high (sometimes even higher) to allow room for a grid of metal pipes, and to permit the lighting director to hang the lights high enough above the scenery to be outside the camera's field of view. The grid is also used to hang scenery, and, if the pipes can be moved individually by offstage controls, to fly set pieces (as in a theater, where set pieces can drop into place from above, and be whisked up into the air as easily).

When the studio is empty, it seems cavernous. Once the set is loaded into the studio, the room somehow seems much smaller. Directors are constantly befuddled by the sheer size of the set (despite the best designer's intentions, it always seems larger in real life than it did in the sketches), and are frequently forced into a situation in which a camera operator must stand hard against the back wall in order to shoot the "beauty shot," which shows the entire set in a single picture.

Most facilities have only one studio, which serves multiple purposes (a talk show in the morning, a commercial taping in the afternoon, news at six and eleven). In these cases, sets are carefully built to occupy opposite corners of the studio, or otherwise arranged in a clever fashion. Just about all studios include a "cyc" as standard equipment: a cyc (which was originally short for "cyclorama," a term borrowed from the theater that is no longer used in television) is a large expanse of wall space, usually white, that gives the impression of an infinite background.

A "soft cyc" is a huge, seamless drape, rather thin, which is stretched taut from ceiling to floor; a series of curved canvas-covered "scoops" ease the transition from the white painted floor to the cyc itself, giving the illusion of a single piece, and, if lit properly, infinite space. A "hard cyc" is a huge wall that runs along one or two sides of a studio, whose bottom is rounded to the floor. When the floor and wall are painted the same color (usually white), and then evenly lit, it is almost impossible to tell where the floor ends and the wall begins. Both hard and soft cycs are usually bathed in colored light, providing an extremely flexible background that's suitable for a wide variety of uses. A lit cyc can be the only set (common when the subject is just a talking head), the backdrop for a bare-bones set

Cyc
A hard cyc provides a sense of infinite space, as floor and walls appear to blend into one another. Note the broad floodlight behind the director (left) and the spotlight, with barn doors open, behind the producer (right). *(Photo courtesy Video Corporation of America)*

(where, for example, a ladder and a director's chair are the only set pieces), or as the unifying background for a large stage setting (as in a local news show, where the set itself is thrown into darkness at the end of the show and seen in silhouette against the light blue [cyc] background). A cyc with minimal set pieces can be extremely effective as a setting for a ballet or other dance performance, because attention is focused on performance, and not on the setting.

All television studios are air-conditioned. A studio has no windows, so air must be circulated artificially. The heat of the lights demands a constant flow of cool air as well. Studio air-conditioning systems are not without their problems, however. Most studios are maintained at temperatures cooler than room temperature, so sweaters and even lightweight jackets are common. The air-conditioning ducts are not always placed in unobtrusive corners; unpredictable air currents cause a drape or a soft cyc to blow in the breeze, and studio drafts have caused many a performer to catch a cold or flu. Some systems are noisy as well, so directors have taken to turning off the air conditioning during the actual taping, turning it on only during breaks and meal periods.

Aside from a few tables, some storage lockers, maybe some audience seating, and three cameras resting in a dark corner, that's the whole story of an empty studio. It's not a very impressive place on first glance, except, of course, to the producer or director who can envision a full-scale production. One day, the studio is a dusty cavern. Two days later, it comes to life. A set is up, lit, and soon the rehearsals will begin.

THE SET

It is the set designer who creates this magic, oftentimes working with a barely adequate

"American Movie Classics" Stage Set

Designer Jonathan Arkin's first concept for American Movie Classics was based on a larger studio space and a more complicated production. As plans were finalized, a more compact, multipurpose setting proved to be the answer—in a smaller studio space. The revised sketch (A) is similar to the finished set (B) with the exception of the jukebox in the center. *(Courtesy Rainbow Program Services)*

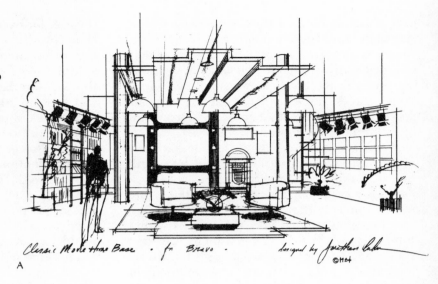

Classic Movie Home Base · for Bravo · designed by Jonathan Arkin
©1984

A

B

budget, using height, depth, width, perspective, pattern, color, and contrast to their greatest advantage on camera. Many times, the set will appear to be "okay" in the studio. But add the proper lighting, and then take a look at the set on camera, in the control room. The camera makes a good set even better.

After seeing a set on camera, most people are surprised to find that the set looks much smaller in real life. The reason lies, most often, in the use of wide-angle lenses for beauty shots. The designer uses these distortions to their greatest advantage, and then creates only what will actually be seen on camera. Pillars, for example, may end just above camera range. Rooms rarely have fourth walls; the audience just assumes that the room is complete. Many sets do not have ceilings, but a well-placed chandelier or skylight that dips into the camera range provides more than an adequate illusion. The sets used on soap operas are the most fascinating; the studio is full of sets for individual rooms, but the area just outside the room is rarely anything more than a painted flat. Take a careful look outside a room window on a soap—and bear in mind that the "view" is nothing more than a painted flat located just behind the glassless window frame.

The set designer, whose role is described in chapter 9, is very much a part of the preproduction and studio production processes. The work begins with a few informal sketches and proceeds through several revisions and approvals. Through it all, the producer and director must depend upon the set designer's talents and abilities to create "magic."

THE ROLES OF THE PRODUCER AND STAFF

For days, weeks, even months, the producer and his or her staff of writers, associate producers, production assistants, and production interns have been busy in preproduction, thinking through every detail, approving, revising, rethinking, getting ready for the show. A panic sets in several days before—"We'll never make it; there's too much to do!"—and, somehow, it all works out. During the preproduction period, the producer and staff must think of everything, including all of the activities in the studio and just about all of the activities related to the talent. As the project heads toward the studio, the staff members begin to specialize. The responsibilities in the office may be all-encompassing; in the studio, the same person's responsibilities may be quite specific. Example: in the office, a PA has been on the phone constantly with a guest, working through every detail of a complicated presentation; in the studio, the same PA might pop her head into the green room and say hello to the guest, and nothing more (because that PA has other responsibilities during production); in the studio, all guests on this particular production look to only one person, the associate producer specifically assigned to the task. The production staff is responsible for keeping the show elements in line, making certain that the graphics are in the proper order, that the guests are made up (in larger markets, and for all network and syndicated shows), understand their segments and are comfortable with what they are going to do, that the crew members have up-to-date scripts, that the cue cards are correct and in the right order, that there are flowers in the dressing room, that lunch has been ordered, that tomorrow's guests are confirmed, that the car is waiting outside to take the vice-president to the airport, that the tapes are being labeled properly. The studio production should never be slowed down or stopped because something is missing, because the PA forgot to check on something, because an associate producer "assumed" it was taken care of. The production staff has plenty to do, and very little of it has anything to do with the cameras, the lights, or any of the technical aspects of the production.

When the rehearsal or the shooting is set to begin, the director takes charge. The director must have complete control over the performers, the technical crew, and the stagehands. They should respond only to the director's voice; the producer's comments should always go through the director, never directly to the talent or to members of the crew. When the producer and the director are working together in the control room, the producer must be careful to keep his or her voice down, because an instruction intended for the director might be heard on headset by a technician and understood as a command. (Example: as producer, I shouted to the director to move in for a tighter shot. The cameraman heard the command, assumed it was from the director, and made a quick zoom to improve the shot, forgetting for the moment that his camera was on the air. The result was a jerky shot, caused by a producer who spoke too loud. The director gave me a dirty look, I apologized later to the cameraman and to the director, and promised it would not happen again.) A good producer will maintain an overview, give the director some necessary feedback, make comments and changes when necessary, and offer approvals and disapprovals in a professional manner.

THE ROLES OF THE DIRECTOR AND THE AD

The director's job, quite simply, is to get the job done. He or she is hired by the producer to take charge in the studio and to deliver a finished product. It is a "buck stops here" job, where there are no excuses, no apologies for a job done poorly. Once the director sets foot in the studio, he or she must be ready to make decisions, to motivate and encourage the best possible performances, frequently under the pressure of a tight schedule.

When the director first sets foot in the studio, he or she should call a crew meeting, and explain the entire project from beginning to end. It is especially important to talk about the production as it will finally be seen by the viewer, to provide the crew with an overall idea of what's happening and why (since so many projects are recorded out of sequence, and crews are not usually exposed to a completed script, it's important to give an overview).

1. The Crew Meeting

The initial crew meeting should also include a technical and staging discussion: a look at the floor plan to see where the set and cameras will be, how the cameras will cover the action, how the microphones will be set up to cover the sound, and so forth. These discussions should be followed by ongoing contact with each of the key crew members (everyone is on a first-name basis; during the setup of shots, directors usually speak to "Diane" or "Pat," not to "camera 1" or "camera 3"), as specifics become clear and changes are made.

It is important for the director to clearly establish the lines of authority during the first production meeting. In every meeting, it is important to remind the crew that every production is a group effort, and that everyone should make contributions and suggestions for the good of the project. A good director will encourage the crew to take an active part, to devise plans for improvement (subject to the director's approval, of course), and will then take the time to listen and to consider each idea carefully. The crew quickly senses a director who sees them as nothing more than operators of equipment ("trained dogs," in the words of one crew member), and quickly grows to resent the rapid-fire one-word commands, the finger snaps that accompany cues, the air of tension and the sense of "no time." The crew is an absolutely essential element of the production, made up of individuals who tend to have considerable experience and a great many ideas. A good director will work the crew hard, earn their respect, and will inevitably see their best creative work besides.

2. Blocking

Once the studio is set and the technical facilities are turned over to the director, a project should be blocked (cameras and performers given their positions, their moves, and their cues), rehearsed, and then produced either live or on tape. Too often, a schedule is written without sufficient time for blocking or for rehearsal, and a director must simply point the cameras and depend upon experience (and luck) for the rest.

The right way to produce a studio production is to allow two to three times as much time for blocking and rehearsal as for the show itself. Even the simplest hour-long talk show should be blocked and rehearsed for about two hours, just to be sure that everything goes as planned.

The time allotted for blocking is the director's time. It is used to check camera shots, to run through some of the more complicated sequences (all preplanned in the director's script) to see whether the shots look good in reality (planning on paper is one thing, seeing the images on the screen may be another), whether the cameras are interfering with one another (either by moving too close to each other or by appearing in another camera's

shot), whether the camera operators can make the kinds of moves that were planned by the director as he or she marked the script. The blocking time is also for the stage manager and stage crew, as they run through key set and prop moves, and make adjustments as necessary ("There's not enough time during the commercial to move that big set piece out of the way; since we're live, we'll have to find a different way to make the move"). Once the blocking of cameras and the staging problems are resolved, performers are brought in (camera blocking is usually done with stand-ins; in network situations, these are union members, but they are usually "anyone in the studio who's free" in most other situations). A performer must be told where to move and when, again making changes in the original plan as performers, camera operators, and the director walk through key sequences. Blocking is also useful for other crew members, like the audio engineers, who will resolve any problems in miking the performers, and the lighting staff, who will make the necessary changes as well. The reasons why changes are made may come from anywhere—a bald spot on the back of a guest's head may be a hot spot when shot from the side (either change the camera or change the lighting), accidental damage to a set piece only minutes before air (change the camera angles to avoid the problem area), the discovery of a need for a second microphone boom when none is available ("I really thought one boom would do the job. We'll have to mike the performers on that side with hand-helds or lavalieres."). Blocking is the time to solve all of these problems; once the rehearsal begins, everything should be working properly.

3. Rehearsal

A rehearsal is a first pass through some or all of a production. There are two types of rehearsals: camera rehearsal and dress rehearsal. The camera rehearsal is mainly a technical rehearsal, when the entire technical operation, the camera operators, the lighting crew, the audio people, the switcher, the video engineer, the videotape ops, are all required to work together for the first time. In this phase, the cues are rehearsed, but the entire script is not (in most cases, the crew members are too busy with their own jobs to hear and understand scripted words, except as cues —"On the word 'elephant,' lights go up on stage left").

A dress rehearsal is primarily for the performers, who may have been rehearsing previously in a large room with tables and chairs instead of actual set pieces, or on the set itself without the benefit of technical facilities. In most cases, the performers have simply read the script and the dress rehearsal provides the first (and only) run-through of the script prior to shooting, and the first opportunity to interact with the cameras. Since the camera rehearsals deal only with cues, the dress rehearsal provides the crew with their first exposure to the project as a complete entity. It is during the dress rehearsal that the production really comes together, and when more problems are solved. With the camera so far back on the opening shot, the host cannot read the teleprompter. The key light is hung too low, and it's blinding a guest. One of the performers "sits funny" and keeps breaking the connection on his wireless microphone transmitter. A guest speaks too softly and the boom can't cover him without picking up every extraneous sound in the studio. A sequence runs too long, it doesn't play, and roughly two pages will have to be rewritten before taping begins. Because of the new camera positions, three of the dancers will not have easy access to backstage, so they won't be able to make the costume change in time. The professor turns her back to the camera every time she writes on the blackboard . . . and she

Rehearsal
In this public service spot for an animal shelter, the "rehearsal" amounted to setting the dogs in their proper positions and making certain that they would stay put. Rehearsals vary in complexity, depending upon the needs of the production. This spot was rehearsed and shot in one morning. Note the use of "seamless paper," available at any photo supply house, as a portable cyc. This is a common setup for commercials, particularly for product shots. *(Photo courtesy Video Corporation of America)*

screeches her chalk besides. Most of the shots look fine, but the AD notices a boom shadow on camera 2's medium shot. The singer cannot hear her tracks, so her lip-sync is off. The list goes on. Every production has its own problems, and all of these must be solved during dress rehearsal (either by stopping in the middle of the rehearsal, or allowing some problem-solving time afterward). Some dress rehearsals are as good or even better than the shoots themselves; for this reason, it may be wise to tape the dress rehearsal (and, in the minds of some directors, not tell a soul, to make certain that nobody gets nervous). Although most of the dress rehearsal is likely to be a little unpolished, certain scenes or parts of scenes may be used in the editing of the show. This technique is common on variety shows, and on Broadway productions taped for television (where the dress rehearsal is taped, and intercut with two or three performances before a live audience and at least one performance that is stopped and started to pick up key close-ups).

Since the needs of every production are different, and budget limitations are most often seen in the amount of available studio time, it is important for the director to assess the amount of time needed for blocking and rehearsal. While there is no excuse for mistakes caused by insufficient rehearsal or blocking, a good director will be able to guide his or her cast and crew with a minimum of preparation. Most directors have learned to think in terms of editing, to record the production in pieces that are later assembled in a polished form. Indeed, very few productions, very few complete speeches, in fact, are seen exactly as they were recorded. With a few audience reaction shots to cover the transitions, even the most disjointed production can be made to look smooth and well produced.

Since editing plays such an important part, even in the simplest of productions, the director, or when the budget permits, the associate director, must take very careful notes during the taping. A complete videotape log must accompany every reel, noting scene numbers, taking numbers, timings, and notes that will be helpful in the editing room, like information on possible cuts, camera errors ("slightly out of focus on the zoom into the house"), performer highlights ("great smile at the end of the take"), or staging problems ("Jimmy started to cry in the middle of her speech"). Tapes must be carefully labeled as well. All of this should be arranged before anything is recorded. Staying organized saves a lot of time.

The director and associate director must be careful to establish very clear communications shorthand for use during the program. Each is connected to a headphone communications system (usually called a "PL" for "private line") to all of the camera operators, the audio engineer, the lighting director, the technical director, the video engineer, other key technical personnel, and the stage manager. Although most directors use fairly traditional terminology for setting shots and cues, most directors have their own way of communicating some specifics. Example: a director who goes by the book and says, "Ready camera three . . . take three" wastes words, and if he's cutting fast, he'll trip over himself. "Hold three . . . take three" uses only four syllables instead of seven or eight (depending on how you pronounce "camera"). "Hold three . . . three" is even faster, but can be confusing for a crew if they're not ready for that kind of brevity. It's important to establish crew communication during rehearsal.

Performers also depend upon cues from the director and from the stage manager, who is the director's representative on the studio floor. Some stage managers (also known as "floor managers") use hand signals for everything (e.g., holding up two fingers to show that two minutes are remaining); others prefer handwritten signals ("two minutes"). If a performer expects signs and sees hand signals, or vice versa, a problem is likely to result. Once again, a few minutes of discussion to standardize cues is essential.

With everyone communicating, and rehearsal set to begin, let's return to the studio floor and have a look around.

THE STUDIO IN ACTION

For days or weeks prior to the start of studio production, arrangements are made; details receive the complete attention of the producer and his or her staff. The engineering staff is informed of equipment requirements and any special technical considerations. The building maintenance staff is apprised of any special situations, such as an audience, live animals, or especially large or expensive props.

Finally, the set is in place, the lighting is just right, the equipment is checked out and

The Director Communicates

Although every director works with variations on the theme, most directors use the same basic commands in the studio, in the control room, and on location. These commands are used to communicate not only with the SM and AD, but with the talent and the technicians as well. Facilities tend to have their own names for certain pieces of equipment, and certain processes as well. Frequently, the process is named for specific equipment, like a Quantel (a device that manipulates images by computer control), and the name remains in use despite the installation of the new equipment.

Before the Production

Set   Putting the set or a set piece into position. (Used as a verb.)

Mark   A spot on the studio floor where a set piece or a prop will sit, or a performer will stand, usually marked by the stage manager with a small piece of masking tape.

Strike   Moving the set out of position, or moving any set piece or prop out of position.

ESU   Electronic setup. Usually seen on production schedules to indicate the time allotted for the engineering staff to set up the cameras, the audio gear, the monitors, and any other studio equipment.

Chip charts, charts, or chips   Test patterns used for camera setup, so that the images are lined up (or "aligned") and the relationship between horizontal and vertical is correct and adjusted for proper rendition of color and of

black-and-white. The word "chip" comes into play because of the appearance of one of the charts. The process is sometimes called "chipping," and is more often called "cameras on charts," and generally occupies ten to fifteen minutes toward the end of ESU.

Camera matching   On a multiple camera production, the images from all three cameras must look alike. On a production where the cameras are poorly matched, one image might be slightly green or orange, one might be slightly brighter than the others. The video engineer works with the lighting director and the camera operators to match the cameras during the camera setup period.

Read-through   A request to the performer (or a stand-in) to read some or all of the script aloud, usually without much movement. A read-through in an office or rehearsal hall is usually an opportunity to hear the script and to make the necessary changes. A read-through in the studio is usually done for the benefit of the director and the technicians, who develop a sense of how the show will play, and how cues will actually fit into the scripted material. The read-through usually provides the director with the first accurate running time for the script, as time is allowed not only for words but for movement as well.

Run-through   A rehearsal of some or all of the script, with some or all of the production elements. Run-throughs are used to give both performers and technicians an opportunity to work together, to work with the cues and staging instructions at least once before the production is taped or goes out live.

The run-through will also provide a more accurate timing. On most projects, particularly those with fixed daily routines, the director runs through only the new or difficult material, and then either reads through or skips past the rest.

In the Control Room

In the control room, the director's attention is trained on four main areas: performance, cutting cameras, monitoring all audio and visual signals, and watching the clock.

1. Communicating with the Studio Floor   Communication with the performers is usually accomplished through the stage manager (see separate sidebar). Prior to production, and during commercial breaks, the director may choose to speak to the performers via a public address system (also called SA for studio address). If the communication is more than a simple "Move over," or "Highlight the word 'new,' " it is best for the director to leave the control room and speak to the performer in person.

2. Communicating with Technicians   Most of the director's time and energy is focused on cameras and on working with the switcher. At the simplest level, the director talks to the cameraperson and composes the shot, gives a warning that the camera is about to be on the air ("Ready camera two"), and then instructs the switcher to put the camera on-line ("Take camera two"). In television, however, nothing is ever simple. While camera 2 is being readied, the shot on camera 1 is on the air (or "on-line"), and camera 3 must be physically moved to get the shot that will follow the one on camera 2. So the director must do several different things at once: compose upcoming shots, make final decisions just before the camera goes on-line, and follow the script by cutting between cameras at appropriate times. On a fast-moving production, the director may not have the time to say much more than "Ready two . . . two . . . Ready three . . . Tighter . . . Three Ready two . . . two . . . Three pull back and left . . . three . . ." and so on.

Add the setup of special effects to this combination, and the need for simple, direct code words becomes even more apparent. Most switching consoles can be preset before a production begins, so the director of a news show faced with positioning a graphic in a window on one side of the screen could say "VTR number two in the box . . . go!" instead of "Three . . . you're hot . . . pan left to make room for the wipe . . . ready to Quantel VTR two in the box wipe . . . take two in the box." What's left unsaid here can cause a problem with a new crew member, however: in this instance, camera 3 knows that it always pans *left* to make room for a box, and camera 1 always pans *right*. A new operator on camera 1 or 3 would need to know whether to pan left or right (which is why rehearsals are so important, particularly when a new crew member joins the team). Most directors and crews develop shorthand after working together for an extended period.

Some directors have their own terms for guiding camerapeople into shots. "A tad more headroom," as you might expect, means "Tilt up just a bit." When a director says, "Sneak in on her," the director probably means "Zoom in, very slowly, so that we can barely see the move." A young director generally picks up a few words from a mentor,

then picks up other shorthand from the technicians.

New directors are sometimes confused by "camera left, camera right, stage left and stage right." These terms are derived from the theater, where a performer looking toward the audience defines "stage left" as his left side, and "stage right" as his right side. From the audience's point of view (and, in the case of television, from the camera's point of view), these directions are reversed, so "stage left" is the same as "camera right" and "stage right" is the same as "camera left." The terms "upstage" and "downstage" also come from the theater, where the stage was once canted toward the audience. The far side of the stage was higher than the near side, so it became known as "upstage"; the lower near side became "downstage."

A director's instruction to turn "clockwise" may be shortened to "clock." Counterclockwise may become "counterclock," but is more often "clock right." This begets "clock left," which causes all sorts of confusion, because left is actually clockwise if you think of the clock with the 12 pointing downstage instead of up. Problems of "Do you mean my right or your right?" still crop up, and are easily solved by referring either to "camera right" or "stage right."

3. Monitoring Audio and Video

Communication with the audio engineer is far less confusing. The audio mixer usually picks up his or her cue from the camera cue. Example: when the director takes camera 2, the talent's camera, and then says "Cue talent" to start the speech, the audio engineer knows to open the talent's microphone. There is no need for a separate cue, unless, of course, something special

is required. During the production, the director does not speak with the audio engineer unless there is a problem ("We can't hear her," "Can you make her sound any better?" "What is that hum?"). The situation is similar in regards to the video engineer.

4. The Clock

Communications regarding time are generally called out by the director or the AD, and then relayed to the floor by the stage manager. "Give her a thirty" means "Give the talent a thirty-second cue"; "We're out in five seconds . . . four . . . three . . . two . . . one . . ." means that the show will be over in five . . . seconds. There are no great mysteries in communicating time, just one hard and fast rule: the director's control room clock is the only one that matters. Clocks on a studio wall may not be as precise; there can be only one time standard, and that comes from the clock that sits just beside the monitor marked PROGRAM on the control room wall. (Clocks with sweep second hands are usually preferable to digital clocks, because the circle helps directors to visualize the show's progress.)

5. Recording for Editing

When a program is being taped for later editing, it is frequently recorded in pieces. Each piece is called a "take." A take that is accepted by the director (or producer, whoever is making the decisions) is called a "buy" or a "buy take." Portions of several takes may be edited together to create one "buy take."

When the day's shooting is done, the director officially dismisses the crew, the talent, and the staff with six words: "Thank you everybody . . . that's a wrap."

Stage Manager Communicates

Although many stage managers and performers have slight variations, the cues listed below are the most common. Almost all of the hand cues can be replaced by cards (usually three- by six-inch pieces of cue card, with cues written or abbreviated in bold black letters), which are generally clear to even the most inexperienced performer. The reason why the hand signals are preferred: they're faster (if the SM's cards get out of order, chaos could result).

These cues exist for one reason: to facilitate communication between the performer and the director. If these cues are not working, develop cues that will. The performer must be able to understand every cue, instantly.

When silence is not required, the stage manager usually shouts the cues for all to hear. The stage manager's position is alongside the camera used to cover the principal talent, always easily seen by the performers without causing them to strain.

Stand by   The production is about to start, or to begin again. When the SM shouts "Stand by," everything should stop, the studio should become very quiet and everyone should be ready for action. (A stand by usually precedes a countdown—a director who issues a stand by and then waits five or ten minutes before starting may be surprised when nobody stands by the next time.)

Countdown   The director, the associate director, and the stage manager all shout time cues, usually at the following intervals: 15 minutes, 10 minutes, 5 minutes, 4, 3, 2, 1 minutes, 45 seconds, 30 seconds, 20, 15, 10, 9, 8, 7, 6, 5, 4, 3, and 2 seconds. The stage manager should start using hand signals as well as shouts from about 15 seconds on. After 2 seconds, there is silence, and hand signals are used exclusively (a spoken cue might be heard accidentally, if, for example, a microphone is open an instant too early). The signal for 15 seconds is a closed fist, facing forward. Seconds from 10 down to 1 are shown by fingers. (First the thumb disappears, then the pinky, then the ring finger, then the middle finger, and finally the pointer).

Cue   (As in "You're ON!) A single finger, pointed at the camera. (The tally light, that little red light above the lens, tells the performer which camera is hot.) A good SM will prepare the performers for camera changes an instant before they occur.

Stretch   There's more time left than anyone had planned, and it's up to the performer to ad-lib, to tell a story or a joke, to spend a little more time with the

working properly, and the studio production schedule begins.

1. The Role of the Stage Manager

There is one person in charge of the studio floor: the stage manager (also called the SM or the floor manager). Connected to the control room by headset, he or she is the voice of the director on the studio floor. Throughout blocking, rehearsals, and the actual production, the stage manager is nominally in charge of all performers and stagehands. The stage manager gives the performers staging instructions from the director, and is the single person to whom performers are responsible when they leave the stage to return to the dressing room, or leave the set to make a phone call. The stage manager also provides the performers with their schedules (or "calls"—the time when the performer is expected to arrive at the studio, the time the performer is expected on stage, at the location, in wardrobe, in makeup, etc.). For the stage crew, the stage manager is the foreman, the person who tells them what to do and when, who supervises the work. The stage manager is the single person who coordinates all studio activities. Although it is not a decision-making position, the responsibilities of studio coordination can be complicated and tricky with a combination of egos, personalities, and union rules sometimes fighting the stage manager's best judgments. It is a more difficult job than it appears to be, particularly on a large production, where two or more SMs may share responsibilities (one backstage to handle the stage crew and performers' en-

guests. The signal looks like the SM is stretching taffy—fingers all touch in a semifist, the fingertips on each hand touch, and the hands are slowly, dramatically pulled away from each other.

Speed-up   Telling the performer to read faster, or to move things along because the show is either dragging or running out of time. In either case, the signal is a single finger, pointed at the performer, making small circles in the air.

Say Bye-bye   The show is about to end. The performer is being told to get off the air (this is usually followed by a countdown, so the performer can time the final speech to the second). The signal is a hand waving good-bye. (There is another signal called a windup that tells the performer to start finishing up; a 1-minute time cue does the job better.)

Cut   The tape stopped; we're going to stop and do something else; we're off the air. The signal is a single finger being drawn across the throat—like a knife. (This is sometimes shouted, but there is always a danger of catching the shout on an otherwise usable take, or hearing the shout on the air.) This signal is also used to tell a performer to finish up (an interview, a monologue), rather like the television equivalent of vaudeville's "hook."

There are other cues, but these are the most common ones. Once again, there is nothing "official" about these cues; common sense will be the SM's best guide. A simple "O.K." hand sign can say a great deal, and so can two outstretched palms that say "stop." Two hands cupped around the backs of the ears will usually get the performer to speak up a bit. Two hands covering the ears (with a bit of pantomime to show pain—but not too much, because the performer shouldn't start to laugh) will get almost anyone to tone down. When hand signals fail completely (this happens even in the best of organizations), a few words speedily written on a cue card can get the message across in no time. Be resourceful—it could save a take.

The stage manager is, in many cases, a substitute for an audience. When no studio audience is available, the stage manager (and the camerapeople and anyone else working the studio floor) frequently offer more comfort than a camera lens. Unless otherwise instructed by the director, the stage manager should be nothing but encouraging, never causing the performer to change course as the result of a facial expression, reaction, or apparent disinterest. This can be a tricky situation, and handling it properly can be infinitely more important than learning every hand cue in the book.

trances and exits; one in front, to handle cues and calls).

The specific responsibilities of the stage manager tend to vary from place to place. In network situations, the job is a classification of the Directors Guild, and is strictly limited to supervision of the stage crew, performers, and communication on behalf of the director. Some local stations require stage managers to handle scenery and consider the supervision of the stage crew to be a technical function, supervised by either a crew chief or a technical director. Other situations require associate directors to stage manage, and limit their activities to the cuing of talent, and still others regard the holding of cue cards as an important part of the stage manager's protected union function (in one case, the SM had to hold the cards because they were seen as part of the

man's responsibilities in cuing the talent—but could not write on them).

The traditional role of the stage manager during the production is in the cuing of talent, either by hand signals or by cards.

2. Cameras

A typical television studio will be equipped with two or three cameras, each mounted on a pedestal, a heavy-duty support designed for fluid movement in almost any direction. Additional cameras may be added to any studio, provided that there are inputs available on the switcher (see the control room section later in this chapter), and that there is enough floor space in the studio to keep cameras out of each other's way.

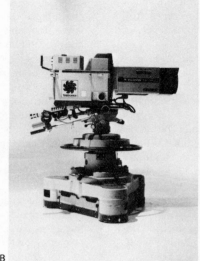

A   B   C

Studio Camera
(A) First, from the operator's point of view. The viewfinder is mounted on top, just above the camera body. The camera is panned and tilted by moving the handles, left and right. The zoom control is mounted on the right handle; a shot box is mounted on the left. The entire assembly is mounted on a pedestal, which permits dollying and trucking. (B) The side view provides a clear view of the camera lens, mounted on the front of the camera body. (C) The shot box allows the operator to preset zooms; pushing a numbered button causes the camera to execute the previously recorded zoom sequence.

The camera operator sees the images taken by his or her camera through a viewfinder. On most studio cameras, this is a four- or five-inch black-and-white screen (measured diagonally). The more sophisticated studio cameras allow the camera operator to switch between several different images on this monitor: the camera's own image, or an image generated in the control room, like a split screen. This function can be important when a director asks a camera operator to find his or her position in a split screen quickly (the split places, for example, the left side of camera 2's shot next to the right side of camera 4's shot, and the exact framing of each shot can only be seen on a monitor that shows the images as a composite). Without this function, the director would either guide the operator until the picture is properly framed, or provide a secondary mon-

itor near the camera. Many camera operators use colored grease pencils to draw the basic layout of any special shots onto the camera's viewfinder during rehearsal, which makes framing easier during the actual production (pencils of several different colors may be used if there are several special shots). The back of the camera also includes some picture controls and one or two headphone jacks for use in the PL system.

The camera operator controls the camera, the lens, and the pedestal by using a series of manual, motorized, and computerized controls. Most of the newer camera models feature a motorized zoom lens, ranging from super-wide-angle to supertelephoto, for a ratio of 15 to 1 or even 25 to 1; the use of special adapters will increase the ratio by a factor of 1.5 or 2. (If terms like "wide angle" and "telephoto" are new to you, pick up a basic book about 35-mm photography—all television directors should spend some time with a good still camera to learn about lenses and about composition.) Focusing is still done by eye, although the automatic focusing controls now common in home video cameras will have an impact in the future. Control of the iris, or aperture (the amount of light passing through the lens), is done either by a built-in computer, by the camera operator, or by a video engineer in the control room. (The preference is still the video engineer, who monitors pictures and makes minor adjustments so that they are perfect, although this job is being eliminated by computer equipment that is adequate for most situations.) "Stopping down" the iris cuts down the amount of light and makes the picture darker; "opening up a stop" incrementally makes the image lighter (once again, "f/stops" are photographic terms—if you're planning to be a director, make sure you understand the functions of lenses and light). Producers should learn as much of the basics of photography and light as possible; the information may be helpful at unexpected times, as when a sudden equipment problem threatens the cancellation of a shoot. By understanding the basics of how a piece of equipment works, it is sometimes possible to work with an engineer and make a "silly suggestion" that breaks the rules and permits the shoot to be completed.

A good engineer, a good operator, these crew members are an essential part of the team. Camera movement requires skill and care. After releasing horizontal and/or vertical locks, the operator can move the camera by using two handles at its base. "Tilting up" points the camera lens upward, increasing the subject's headroom and eliminating the bottom of the picture. "Tilting down" is the opposite: the top part of the picture is eliminated instead. "Panning left" points the lens to the left and eliminates the portion of the picture on the right, while "panning right" points the lens to the right, eliminating the picture on the left. Try this with your own 35-mm camera (or pocket camera, or cardboard toweling tube) and you'll see the principle at work.

The zoom is controlled by a lever mounted on one of the camera's handles, which in turn controls the motorized zoom. "Zooming in" increases image size, and "zooming out" reduces the size of the image. Zoom lenses are designed to hold their focus while zooming in or zooming out. When a camera is moved physically, however, the lens must be refocused.

The pedestal allows the camera head to be moved up ("ped up") or down ("ped down") to change the angle on the subject.

Physically rolling the camera from side to side is called "trucking." Moving the camera and pedestal to the left ("truck left") or the right ("truck right") can eliminate problems with "keystoning" (where an object should look rectangular, but because of the camera angle, the left side looks larger than the right or vice versa). A camera also trucks to follow a performer who is moving laterally. Movement upstage (toward the back of the stage) and downstage (toward the front) is called "dollying." To dolly in is to make the move upstage; to dolly back is to move downstage, away from the action. Dollying cameras and pedestals is becoming a lost art; it's easier to zoom than to move the whole camera. But the results are visibly different. When a camera zooms in, it compresses the space between the subject and the background. When the camera is dollied in, the space between the subject and the background will look normal. If you have access to a television camera (even a home video camera or a 35-mm still camera will do), make a few comparisons. Once you understand the processes as a director, you will be able to make the appropriate choices for each situation. A "zoom in," for example, might make a set look smaller and more intimate; a "dolly

The Technician Communicates

Many engineers are in the habit of calling specific pieces of equipment by their model numbers (a small lavaliere or clip microphone called the Sony ECM-50 becomes a "50"). Although there are a great many products currently in use in television facilities, this list covers the most common ones. A half-hour with a friendly engineer will provide a director with the current complement of available equipment at any facility; an hour will provide sufficient time for an industry-wide overview.

Knowledge of specific equipment by number is sometimes useful (particularly when a similar piece failed at a crucial moment on a previous project). In most cases, the equipment selected by the technical staff and their supervisors will be the right equipment for the facility and for the job. It is wise for a director to request an equipment list if unfamiliar with the facility or its people; changes in the basic equipment list are usually handled by renting equipment from other facilities or from supply houses.

Many directors (and producers) find this kind of information useful for browsing and for reference, but, frankly, prefer to spend time on other aspects of the production. A director who chooses the right crew can work well without firsthand knowledge of these specifics. In situations where the director (or producer) is required to do more than is normally expected, however, knowledge of these specifics can be the difference between, for example, adequate and inadequate sound coverage.

1. Microphones   Basically, there are three types of microphones, and each is designed for a specific purpose. A microphone that is clipped to a performer's lapel, or hung around the neck, must be small and unobtrusive. A hand-held microphone must be large enough to hold comfortably, sufficiently sensitive to cover the speaker at a distance of a foot or more, and durable, to withstand some mishandling. A third variety of microphone is required to pick up sounds that are located several feet or more from the location of the mike.

In most studio and remote setups, Sony's ECM-50 (or the improved ECM-250) is extremely popular. It's small, attaches easily via clip or lavaliere, and is perfect for interviews, newscasts, and, incidentally, for music. The Shure SM11 is also popular for use as a clip or lavaliere microphone. These microphones are omnidirectional; their pickup pattern is even in every direction.

There are a great many different desk and hand microphones currently in use, and each has its adherents. Shure and Electro-Voice products are popular; you're likely to encounter Shure's SM 58, SM61, and SM81, and Electro-Voice's 635A, RE-50, CS-15, RE-15, and RE-16, as well as products from AKG, Beyer, Sony, and others. These microphones are all very rugged and all very versatile. Preferences in their sound, their capability, and their dependability are detected only with regular use. All of these microphones have cardiod pickups, with dead zones toward the bottom of the microphone.

Microphones used on booms or as telescopes are designed specifically for long-range, highly targeted pickup patterns (called "hypercardiod"). Sennheiser's 415, 815, and their newer counterparts, the 416 and 816, are popular, as are the AKG 900E and the Electro-Voice DL-42. These are fairly specialized products. Most facilities have one or two of them (as opposed to a dozen or more of other models).

Microphones used for high-quality audio recording are infrequently seen in television setups, except in special cases. The reason lies in the microphone's design: a microphone of superior quality is rarely designed to take the abuse that is common in even the most careful television studios. On music programs, microphones made by Neumann (the U-87 and the larger U-89) are used, as are AKG's 414B, along with a variety of other microphones that are selected by a knowledgeable audio engineer.

2. Lights   Simply speaking, the director is really only concerned about one question as to lighting: Are the performers (the set, etc.) being seen properly? Is there enough light on the key subjects? Are certain areas too dark? Does the whole set look too dark, too light, or well balanced? The director can judge only by what he (she) sees on the monitor, and must provide both practical and instinctual responses for the lighting director.

Once again, decisions about actual equipment are best left to a skilled lighting designer or lighting director. Still, a director should understand the basic equipment available, and what it can do. Different lights are used to create different effects: some are soft, some are harsh, some are designed to provide general illumination and others are used to highlight.

The job of the lighting director is filled with complexities, involving not only the use of instruments, but the (sometimes unpredictable) effect of their illumination on the subject and background areas. The LD must be concerned not only with the technical ramifications of the work (the ratio of the brightest to the darkest spot on camera, for example, can be no greater than 30 to 1), but also the aesthetic (the work must look good on camera). Ultimately, both of these situations depend upon the skillful use of equipment.

All studio lights fall into one of two categories: spotlights or floodlights.

A. Spotlights The most popular spots are Fresnels because of their extreme flexibility. At the front of the instrument is a lens. When the lamp inside is positioned close to the lens, the Fresnel provides a wide, floodlightlike beam. When the lamp is pulled away from the lens, the beam becomes focused, like a spotlight. Fresnels are usually identified either by their diameters (in inches, with the largest diameters representing the largest and most powerful lights), or by their wattages (1,000 and 2,000 watts are the most common, but 5,000-watt lamps are available).

For a beam that can be focused even more precisely, an "ellipsoidal" spotlight is used. These are commonly used to project patterns on a cyc, to create small circles of light (colored light, with a gel) on a studio floor, and for similar effects. Ellipsoidal spots are usually 750-watt instruments, with brighter varieties available from rental houses (but rarely standard equipment in facilities).

The "follow spot" is common in the theater, and is used in television when a similar effect is desired. You'll see several follow spots used on shows like "The Academy Awards" (and in movies like *The Jolson Story*). They are not usually found in television studios, may be rented, and may be standard equipment when the studio is a renovated theater.

B. Floodlights There are three basic types of floodlights. The most common is the "scoop," with its 1,000- and 1,500-watt bulbs. Most lighting grids are filled with scoop lights, because they provide the basic illumination for the entire set. Some scoops have adjustable beams, but most are basic, straightforward lighting instruments (in most facilities, they were purchased when the studio was built and never replaced—they're not fancy, so they last forever).

Several varieties of "broad lights" are used for illumination at the five- to ten-foot level (which is oftentimes essential for the lighting of a product or a person on a commercial). The most common type of broad light is about seven feet tall, sits on a stand, and has a row of bright (and hot!) lights at the base of a box with a curved, highly reflective back. Other broad lights are simply rows of lights. Broad lights are common when control over highlights and shadows is crucial, as on a commercial. Broad lights are not very common in studio situations where people must talk for an extended period, because they're hot and they make the performers uncomfortable (even with plenty of studio air conditioning).

Cyc lights are strips of lights that are placed at the base of the cyc, usually between the vertical cyc itself and the horizontal-to-vertical scoop at its base. Cyc lights are usually used with gels to color the cyc from the bottom; lighting instruments hung from the grid are used to match the color and create an even effect.

There are several tools that are commonly used by LDs to increase control. Most lights are hung from the grid (actually, from the pipes on the grid which are called battens) with C-clamps. In some cases, it will be desirable to position the light nearer to the performer, nearer to the floor (and, yet, because the show has an audience, the producer will not permit any lighting instruments on the floor). The answer is a "pantograph," an accordionlike device that permits any studio light to be pulled away from the grid (as far as twelve feet!). Once the pantograph is installed between the light and the batten, the light can be easily repositioned in seconds.

When lights can be placed on the floor, there are a variety of floor stands available. Lights can also be attached to the front of cameras, clamped to the tops and sides of set pieces, even suspended from special booms. In some cases, lights are placed at angles to the area that must be illuminated, and a reflector delivers the light to the desired area. In other cases, an opaque sheet is clamped near the light to prevent the beam from reaching a specific area.

3. Cameras In communicating with the technical world, cameras present only a few challenges. There are only a few manufacturers—Ampex, Sony, RCA, Ikegami, and Hitachi are among the most popular—and their products are all quite expensive. Most engineers have only worked with one or two of these, and, to a great extent, they are similar. Few directors have the opportunity to select the cameras that will be used on a particular project; they are generally provided as part of a studio package or part of a complement of location equipment. In the major cities, free-lance camera people own their cameras (in these cases, the Ikegami 79A and 79D seem to be the preferred choices, for durability, and in the case of the 79D, for its clarity in low-light situations). Although there are many different products available, lenses are usually purchased with the cameras, and so there is little choice involved, except where special long lenses are needed (e.g., the camera is going to be used to cover a space shuttle launch, and must be placed a mile away from the subject to avoid the intense heat).

---

in" might make the same set look larger, more impressive, and more dramatic.

The wheels on a camera dolly can be used in two different ways. With all three wheels parallel, the camera will move in a straight line. With one wheel angled, the camera will move like a tricycle, along an arc. The switch is made easily and quickly on any camera pedestal.

When a camera is moved skillfully, the viewer's sense of "being there" is enhanced. Note, however, that some studios limit the movement of cameras because their floors are not in good shape. Rolling over chips or bumps in a studio floor can really cause the picture to shake; cigarette butts can have a similar effect. Fluid camera movement also requires a skilled and experienced operator.

Lightweight cameras designed for location work can be used in the studio as well, either on a pedestal or tripod, or, more often, as hand-held cameras that can move where a pedestal cannot. Typically, a hand-held camera is operated by two people, one who handles the camera and the other who guides the cable. Hand-held cameras can move quickly, can provide unusual angles and intimate close-ups (each move, however, requires not only recomposing the picture but refocusing as well). They are commonly used during rock concerts and staged events like the "Miss America Pageant" to get close to the performers. In the studio, a hand-held camera can be used as an extra camera to pick up unusual shots for variety, for color, for detail. As a director, it is important to remember that the hand-held camera operator will probably move around a lot, so it can be tough to keep him out of other camera shots. With some coordination, this can be done effectively. Hand-held cameras should always be considered to be extra cameras; a good hand-held camera operator will follow the action and offer the director an interesting variety of shots without very much instruction. The director must provide feedback, tell the camera operator when

Hand-Held Camera in Studio
A hand-held camera provides enormous flexibility, especially in studio settings. A good hand-held camera operator can shoot continuously while changing camera height and camera angles, providing some unusual points of view that would be unattainable with a standard camera mount. *(Photo courtesy Reeves Teletape)*

to stay in position and when to move on, when to zoom in and when to try an on-camera move (e.g., starting the camera near the floor and then slowly, smoothly, lifting it while maintaining both focus and composition). On a complicated production, an associate director may direct the hand-held camera(s), offering the director the right shot at the right moment. A note of caution, however: some hand-held camerapeople will go anywhere, climb anything, suspend themselves from anything in order to "get a great shot." Some shots do involve measured risks, but the operator's safety must always be considered first when developing potentially dangerous shots.

A camera can also be mounted on a crane, a large mechanical system that resembles an outstretched arm with a camera on its palm. A camera on a crane can provide some unusual shots, and can do the job of several cameras because of the range of its movement. The crane camera can move up and back to get a beauty shot from a high angle, and can then be used like a standard pedestal camera to cover action on stage, and still be able to move almost straight up for the closing shot in the sequence. All of this movement requires some skill, especially if the moves are to be used on the air (as opposed to placing the camera in a position, then taking the shot). Camera cranes are usually quite massive (the larger the crane, the greater the reach, the greater the number of available shots). They usually require one or two operators in addition to the cameraperson (one drives, the other guides, and guards against accidents).

Smaller cranes, requiring only one driver/guide/guard, are also available. Cranes are usually reserved for bigger-budget talk and variety shows, and for game shows. Small cranes are common in some production facilities that specialize in commercials.

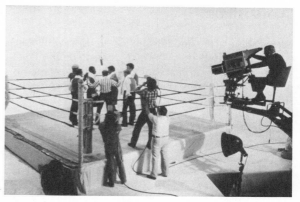

Camera on Crane in Studio
A camera on a crane provides the high-angle shots that give the audience a real sense of place, but it also provides a flexibility of movement. The camera can move sideways at any height, and can work down very low (less than two feet from the ground). A driver controls the motion of the crane from floor level (he's located just out of frame to the right). *(Photo courtesy Reeves Teletape)*

3. Lighting

The lighting director is responsible for the design and execution of the lighting for the production. Since the lights must be hung on the grid before any scenery is in place, the LD and staff do most of their work before the other technicians arrive. It is not unusual for the lighting crew to arrive at the studio at three or four in the morning, when everyone else arrives at eight. The studio must be clear to set the lights.

The LD starts planning soon after the set designer makes a rendering and floor plan available. The lighting plan is based on the designer's work, on the placement of the set in the studio, and on comments from the director about the places where the performers are likely to move. Once this information has been "finalized" (the word is in quotation marks because changes will be made until the very last minute—the LD must be informed of all

Composition

Composition is the arrangement of elements within an image that is pleasing to the eye. "Good composition" is largely a matter of personal taste, but there are some standard rules that are observed by most camerapeople and by most directors. Since every shot is a collaboration between the cameraperson and the director, it is important to agree on a basic style during the early stages of a production (this is very much the case in music video production, where rules of composition are oftentimes violated intentionally to create a desired effect).

Keep It Simple . . .

Good television composition requires an understanding of how the medium interacts with the viewer. Television tends to be an intimate medium, most often watched by one or two people per set. The medium returns this intimacy with a good many close-ups, which help to establish a relationship with the viewer, and with a minimum of wide shots, where performers and set pieces can be too small to be identified. Because the screen is so small, only two or three visual elements can be used to create a picture (more than three becomes confusing, and the images become insignificant). One picture element is almost always best.

In the control room (or editing room), monitors can be adjusted to show the complete image (used mainly by engineers to make certain that the picture is electronically correct) or "TV cutoff" (roughly 40 percent of the picture is lost in transmission—10 percent from the top, 10 percent from the bottom, 10 percent from the left, and 10 percent from the right). It is important to compose shots using a monitor adjusted for "TV cutoff," and to generally avoid placing anything important near the edges of the screen (because TV cutoff varies on each set).

. . . But Not Too Simple

The "safest" picture composition for television shows one subject, size large, in the middle of the screen. Unfortunately, this kind of composition becomes dull after a half-minute or so, and viewers tend to glaze over, losing the connection with what's being said and why.

Equally safe is the two-shot, which shows a balanced pair of subjects. Many television shows are "directed" (the term is used loosely) by cutting between the dependable close-up and the equally dependable two-shot. New directors lean on these two shots while learning how to control the rest of the production. Old directors who lean on these shots are either uninspired or lazy.

The third safe shot is the long shot which shows the set and the performers, with the whole scene framed like a painting or photograph. This is the shot that usually opens and closes a simple production.

Adding Some Imaginative Touches

Placing a subject off-center may not work in most cases, but it is worth trying during rehearsal. In some cases, a subject placed off-center, perhaps in a slight profile instead of head-on, can give the impression of activity, of conversation, of other peole around, of a sense of place. (Just be sure that the eyes face the open area of the screen, and not the edge.) When a subject is basically horizontal, a vertical highlight is oftentimes more effective if placed on one side or the other, and not in the center.

Changing the camera's point of view can add interest as well. Car chases shot from low angles that show screeching tires in the foreground, for example, are far more exciting than those shot at normal eye level.

Television is best with foreground elements (the intimacy factor once again), but foreground pieces can be used to frame background pieces, and to make interesting transitions. A flower in the foreground, in sharp focus, commands attention until the point of focus is changed to the background (this is called "racking focus"); this technique is frequently used on shows like "Charlie's Angels".

Seeking out geometric patterns can also be fun for the cameraperson and for the viewer. If picture elements suggest a triangle, a pyramid, a cross, these can be used to add effect to certain sequences. A trained eye will find these images (and most camerapeople, given the opportunity, will generally come up with some very interesting ideas).

Five Commandments of Composition

1. Watch the natural cut-off lines! Never end the picture at the chin, the bust, the waist, the hemline, the knees, or the ankles. The most pleasing pictures will be framed with the bottom of the picture anywhere in between these natural cut-off lines.
2. Always lead a moving subject, never follow it! When a subject is walking, running, driving, or moving across the screen in any way, make sure that the subject is kept slightly off-center most of the time, leaving room in the viewer's mind for movement in the forward direction. Varying the exact placement of the subject on the screen is one way to show that it's in motion.
3. Watch the backgrounds! Avoid placements of subject and camera that result in trees, posts, plants, and even straight lines growing out of a subject's head. (Trucking slightly to the right or left will usually solve the problem.)
4. If you run into problems, keep the camera moving! Sometimes, a situation cannot be controlled. One way to avoid problems in composition is not to stay on the problematic pictures for too long. Zoom in or out slowly to change the framing and distract the viewer's attention from the problem; truck or dolly for the same effect. And, when all else fails, intersperse the sequence with background shots to cover the composition problems on the master shot.
5. Let children and pets dominate! Television loves kids and animals, and no attempt should be made to draw attention from them. Adults who appear in the same picture as kids or pets must kneel down to their level (never bend—it looks strange and unnatural on camera).

Adding Even More Impact

Due to optical illusions caused by lens design, impact can be given to, or taken away from, subjects by varying camera angles. Shooting up at a subject, for example, tends to make the subject look larger and more formidable. Shooting down tends to diminish the subject's importance (an interesting technique that might have been used by an unscrupulous director who supported a male candidate and wanted a female opponent to look weak).

Action moving toward the viewer tends to have more impact than motion across the screen. Watch any commercial for a sporty car and take note of the number of head-on action shots. Most are wide-angle shots, incidentally, because a wide-angle lens causes the viewer to imagine faster, more powerful action, while a telephoto shot tends to slow things down.

Experimentation Is Important

One of the ways to keep a show fresh is to experiment with different camera angles and different styles of composition. Camerapeople should be encouraged to try new ideas, provided that there is sufficient time to do so (e.g., time is not taken away from the director's rehearsal or blocking schedule).

When a television camera is not available, 35-mm photography provides an excellent, and comparatively inexpensive, tool for experimentation with the effects of lenses and practice in good composition techniques. Many directors and camerapeople had been amateur still photographers for years before entering television.

---

changes), the LD designs a lighting plan that details the position of every light in the studio, its power, the type of instrument to be used and its function. Lights are then hung, principally on the grid, and on stands and even on the edges of the set pieces in certain situations. (Sometimes the lights are hung before the set is put into place, but the opposite is more often the case.) The lighting is then fine-tuned, as the LD checks for unwanted shadows (from booms, from cameras on crane, from performers crossing in front of a light), hot spots and areas that are not sufficiently bright (checked by eye and with a light meter). When cameras are available, the lighting director makes the final changes, and then shows the completed work to the director, who may suggest changes at any time. On-camera blocking and rehearsal provide the LD with an opportunity to see the results of the lighting plan and to revise accordingly. An experienced LD knows that changes are inevitable, and must know how to work quickly, how to make a few small changes that will correct a problem

LD under Grid
The LD and his crew set up the studio for a children's talk show, with audience onstage. The four scoops will provide most of the illumination for the left bleacher. Space must be allowed between the bleacher and the wall for backlighting, so that members of the audience will look three-dimensional.

without requiring the handling of too many instruments. It is what's on the screen that counts; good lighting is always judged by the result seen on the monitor, which may "break the rules" of standard practice or accepted technique.

During the actual production, the LD sits behind a console, with faders that control the intensity of individual lights or, more often, groups of lights. If there are lighting changes required during the shoot, these "lighting cues" can be accomplished by either moving the faders by hand (on a manual board) or by calling up a cue by number and letting the board do a nice smooth fade (on a computerized board).

The mechanics of lighting are best left to the LD. It is one of the few areas in television production where knowledge of individual pieces of equipment is not essential, unless you have a particular interest in lighting design. As a director, it is useful to know the difference between a "scoop" (which spreads light over a wide area) and a "spot" (which focuses the beam on a specific area), and a

"cyc light" (which is used to light the cyc evenly, and to bathe it in color). You should be familiar with "gels" (colored sheets of gelatin that are placed in front of individual lighting instruments to color the beam), and the various diffusion and reflection systems described in an earlier chapter. Although the director will make suggestions based on what is seen on the television screen, it is helpful to know the basics of lighting design as well. A subject is lit first by a "key light," which is hung in front of the subject, just slightly off-center to create subtle shadows. A "back light" highlights the hair, separates the subject from the background, and generally makes the subject appear three-dimensional. It is hung just behind the subject, usually centered. One or more "kickers" or "side lights" are used to complete the image, to add more subtle shadows (this shadowing process is called "modeling"; the word is also used as a noun). "Fill light" washes the entire scene with light, so it is helpful in darker areas of the set. Fill light is not focused light, so it may disturb earlier attempts at modeling. The bal-

LD at Console
After hanging and focusing the lights, lighting director Michael Pelech sets up the board for light cues. Each instrument, or group of instruments, is on a numbered dimmer. Once the audience fills the bleachers, he will look at the picture on the monitor and make the necessary adjustments.

ance of these basic techniques—so many of them are used over and over again when lighting a complete set—is the lighting director's most difficult task. When one light is changed, the new scheme may affect several groups of lights. This is why most lighting directors insist on up-to-date information about the set and the staging, and why there may be reluctance to change lighting schemes once the lights are hung and focused.

4. Cue Cards and TelePrompTers

Most television projects are produced on relatively short schedules, so performers rarely have the time to memorize scripts. Basic familiarity with the script is all that most performers require, because every word will be available during the actual performance, clearly printed either on cue cards or on a TelePrompTer (or "prompter").

Cue cards are large sheets of white cardboard, heavy enough to stand upright without flopping over. The cards are held just to the left or right of the camera lens, so the performer looks as though he or she is looking at the lens and not at the cards. (A note of caution here: a performer will appear to be looking at the cards, not the camera, if the camera is closer than about eight feet from the performer). A thick-tipped black marker (with a red one for emphasis) is used to write the cue cards by hand. On some projects, every word is printed on cue cards; more often, however, only key words are carded (a TelePrompTer is used if more than key words are required). Cue cards are ideal for short pieces, like commercials or comedy routines (where Bob Hope, for example, sees one joke on each card). Cue cards are not ideal for fully scripted work, but many actors, variety and comedy performers prefer them because the letters are big and easy to see.

TelePrompTer in Use
A TelePrompTer can be used in the studio or in the field. Here, the prompter (a scroll-type machine, operated by a motor) is mounted above the camera lens. The top mirror reflects the image onto the bottom one, which is physically in front of the lens. The talent reads the script and appears to be looking directly into the lens. The lens does not show the prompter's image, because it cannot normally focus on something so near. Operator Dean Francis can change the speed of the scrolling with a hand-held controller, so that the script never lags behind (or speeds ahead of) the reader.

One step beyond the cue card is the scroll-type TelePrompTer. Here, the entire script is typed in inch-high letters and then scrolled, either by hand crank or by motorized control, just above or below the camera lens. With a series of mirrors, the type can be made to appear in front of the lens itself (the image is too close to the lens to be seen on camera, but it is very clear to the performer, who appears to be looking directly into the lens as he or she reads).

The most popular prompting system involves a conveyor belt, a second television camera, a monitor, and a series of mirrors. Here's how it works: a specially typed script is fed, page by page, on a conveyor belt, and is seen by a small television camera. This television signal is fed to a black-and-white monitor that's mounted just above the performer's

Cue Card and TelePrompTer Tips

Cue cards and TelePrompTers can be useful tools, but they can cause problems if they are used improperly. Larger productions generally hire specialists who take care of the cards or the prompter. Most productions do not have the budget (the money is better spent on an additional camera or an extra set piece), so production people take the responsibility for these prompting systems. The tips that follow should eliminate some problems and make the whole system efficient.

1. Cue cards should be white on both sides, of medium weight, and either twenty by thirty inches (larger cards fit lots of copy, but they can be hard to handle if you're short), or twenty by fifteen (easy to handle, but they only fit a sentence or two). The cards are white on both sides so that each card can be used twice; just be sure to make a big X on the side that's been used, because cards do get flipped by accident.

2. Every cue card should be numbered, first with the scene number and then with the card number in sequence ("12-6" would be scene 12, sixth card). Numbering should be done with a ballpoint or felt-tip pen, preferably in red, in the upper-left corner of each card, and should not be large enough to be read by the talent.

3. Markers should be broad-tipped, black and red. The black marker is used to write the script (ALWAYS IN CAPITAL LETTERS). The red marker is used for instructions (emphasis on a word or phrase, "hold up product," etc.).

4. Letters should be large enough to be seen at ten feet (and if the performer is nearsighted, be sure to do a test before writing the whole script on cards with letters that are too small). Two-inch letters are about normal. Spacing should be even, and letters should not be crowded at the end of a line. Test the letter size with the performers before you write the cards.

5. If a mistake is made, do not show the cross-out. Instead, cover the mistake with two-inch-wide white masking tape (available in most art supply stores, along with the cards and the markers) and write the correction directly on the masking tape.

6. When using punctuation or other special marks, make certain that the performer understands the symbols. Most people recognize an ellipsis (" . . .") as a pause or as an expectation that something will continue, for example. Asterisks have different meanings (a shrug to one performer, a smile to another). When a sentence ends, make the period larger than normal to be sure it is seen, so that the performer does not run one sentence into the next ("ORSON WELLES STARRED IN CITIZEN KANE BECAUSE OF A SINUS CONDITION . . ." instead of "ORSON WELLES STARRED IN CITIZEN KANE. BECAUSE OF A SINUS CONDITION, THE START OF SHOOTING WAS DELAYED THREE WEEKS"). Difficult words and foreign phrases should be spelled phonetically, again, according to the performer's own codes ("Ixtacihuatal, the extinct volcano north of Popocatepetl" becomes "IX-TA-SEE-WHAT-UL . . . THE EXTINCT VOLCANO NORTH OF PO-PO-CAT-UH-PET-UL"). Highlight or underline syllables for emphasis.

7. When working with several cameras, a single set of cards may be limiting (the performer can only speak into the camera with the cards next to it). A matched set of cards for each camera, each with a person to pull them, is one solution. TelePrompTers on each camera are a better solution.

8. Most of the cue card tips apply, with some adaptation, to TelePrompTers. Clarity is important, even when handwritten changes must be made on masking tape that covers the large-typed copy (on a scroll-type system), or on correction tape placed over the error on script pages (on a conveyor belt system). All of the tips on punctuation, special markings, and pronunciation apply.

camera. A series of mirrors reflects the image of the script just in front of the lens, as in the scroll-type system. Since this system is based on script pages, it is a favorite of news programs, where last-minute changes and rewritten script pages are very common. No special typewriters, no cumbersome scrolls, no heavy cue cards are involved. Just paper, marked with a special column where the copy should appear (usually on the left side of the page).

The key to any of these prompting systems is pacing. Any one of these can be effective, provided the operator (or, in the case of cards, the holder) keeps up with the performer, never moves too quickly ahead, or too slowly. The prompting systems can be adjusted to move more quickly, or more slowly. Talent should always have a script within easy reach, with pages turned when possible, just in case the prompting system fails, or the cue cards fall over. Each card change (in the lingo, cards are "pulled") should occur just before the last words on the card. Expect the unexpected; accidents do happen, and they are not easily undone when the performer is working live, or in front of an audience.

One last word of caution: when working on a production where script changes are coming quickly, a production assistant should be assigned to working with the cue card or prompter people, feeding them fast and accurate changes, and double-checking EVERYTHING, including the order of the cards and the script, just to be sure. (Cue cards and script pages should be numbered, with letters for revisions; establish a common code to save time.)

5. Monitors and Other Studio Gear

Every studio should have one or more monitors (e.g., quality television sets without tuners and without audio circuitry or speakers), connected directly to the control room to show the program as it is being recorded or broadcast live. In most cases, monitors should be faced away from those appearing on camera (as in the case of a studio audience who will look at themselves on TV, and not at the cameras). Even the most experienced performers will be affected by the way they look on camera, and by the way the whole production looks (since the performer is key to the entire presentation, many feel free to express opinions about camera angles, lighting, and script —if a performer is a good producer, this can be a pleasure; in an effort to "help," a performer may become involved in areas outside performance, without necessarily understanding the bigger picture). Monitors are also to see special feeds; a newscaster, for example, might not see the program feed, but must be able to see the videotape playbacks. Sound is monitored through speakers that are controlled by the audio engineer. These are not used during the shoot, because microphones will either cause feedback or a boom-y effect. The sound system is used during rehearsals and during videotape playbacks, and may be used either to feed program material or as a public address system for the director and other control room personnel. Audio and video monitors are usually suspended from the ceiling or the lighting grid; additional units may, of course, be placed on a floor stand, a table, or elsewhere in the studio.

There are other sights that are common to most television studios, including fire extinguishers, prop lockers (the equivalent of a grip kit on location), some audience seating, a few extra flats (usually left over from other productions, but usable after a paint job), extra cables and extension cords, extra lights, and, most important of all, a table with coffee and Danish (bagels, in New York). Ashtrays are not usually found in television studios, because of the potential for fire and because cigarette butts on the floor can cause a dollying or trucking camera to bump, or to stop completely. Cigarettes left burning on tables or furniture (a terrible habit, but one that's fairly common) can cause burn marks, which cost money to fix. Coffee is best kept in a special area (preferably outside the studio, in a separate room), because it can stain the set, and because it (or any other liquid) might accidentally spill into or onto a

Control Room
The "Entertainment Tonight" control room, in action. The director is in the center, facing the bank of monitors. The PROGRAM and PRESET are the large ones, in the center. The four camera monitors are along the bottom left. The technical director sits to the director's right, with the large switcher console before him. The producer is on the director's left, back to us, following the script. *(Photo copyright © 1985 Paramount Pictures Corporation. All rights reserved)*

piece of sensitive electronic equipment, rendering it useless (the repairs are usually very costly). Rules should be established from the start: nobody on the set except the performers (talk show interview chairs are not intended to be resting places for a tired PA or member of the crew), no personal belongings on the set (they might be left, accidentally, and appear on the air), no liquids on the set (except water for the performers), and no smoking. A producer who fails to enforce these rules in the studio and the control room is inviting trouble.

THE CONTROL ROOM

To a television director, the control room is home. The director sits at the center of a long console, looking up at a bank of black-and-white and color monitors. The director is in control of all of the images on those monitors, responsible for composing and selecting the best possible pictures for use in the completed production. Since this selection is usually done in real time (as the events are happening, editing plans notwithstanding) and frequently requires a combination of several of these images as well as coordination with audio and other production elements, the director should be given ideal circumstances for concentration. During the production, NOBODY speaks up in a control room, unless they are giving cues (whispering is okay, but best kept to a minimum). The busy control room is not a place for an account executive to impress a client, or for a company president to drop in on the staff (I have personally asked some of the best people to either shut up or leave—I suggest you adopt the policy). The control room is like the cockpit of a plane in flight, operating in an atmosphere of controlled tension and extreme pressure, where the slightest distraction could result in a serious mistake. In most cases, the director must guide and pace the action, and the coverage of it, and then make split second decisions which are inevitably based on plans that changed seven times in the last hour. Directing is a tough job, harder than it looks because there are so many details that must be kept in a precisely timed balance. Most directors thrive on the tension, the energy, the excitement, the exhilaration, the control. Directing requires a special type of character who can maintain a calm while operating at optimum speed and efficiency.

When a director directs, he or she usually works from a combination of a marked script, observation of the way that the show "plays," and instinct. The script contains the basic cues: when to roll videotape, when to cut from one performer to another. But these decisions are best made by watching the action, and going with the flow, helping to control the pace when necessary by offering a speed-up or a stretch, in order to follow the requirements of the format (and to fit in the commercials). The director is constantly looking ahead, anticipating the next three or four moves, providing advance notice to the crew members whenever possible. When one camera is on, another camera's shot should be ready, and another should be finding its next shot. The basic rules of location work apply: first, the establishment shot, followed by a series of close-ups, with occasional medium shots and two-shots to remind the audience of time and place. Television is a close-up medium where faces count for a great deal of the overall impact of any production; close-ups should, therefore, be the most-used shots. At least one of the cameras should be on a close-up of the principal performer (not too close—a headshot will do for most situations), or on the principal action, at all times. The other cameras should be used for support and illustration, and, whenever possible, should be used to offer some variety: the occasional profile shot, a close-up of someone in the audience, a reaction shot, a two-shot that shows the relationship between the interviewer and her subject. Orchestrating the cameras is something that can be learned by doing plenty of planning, working out floor plans and marking the script to define every camera shot in advance (at least in concept), and gaining experience whenever possible—you will never learn it by reading a book. However, you can learn a great deal by watching a director at work in a control room, and asking questions after the production is over.

1. The Switcher

As the director is making the camera decisions, the "switcher" (or, on larger productions, the technical director or TD, who also takes charge of the crew and is responsible for all technical aspects of the production) is the one who actually makes the director's instructions come to life. The switcher responds to all of the director's cues, cutting, dissolving from one image to another, adding special effects on cue. The switcher operates a control board with as many as several hundred buttons, switches, and levers. The board is laid out in a series of "buses"—lines of buttons that are used to "take" any camera, any videotape playback, any remote feed, any special-effects device or an input from another bus (see below). The labels on the buttons correspond to monitors on the control room wall: camera 1, camera 2, camera 3, camera 4, videotape 1, videotape 2, videotape 3, videotape 4, remote feed 1, remote feed 2, and so forth, depending on what incoming feeds are connected to the board and to the monitors above it. Each of the buses is nearly identical. The most important of these buses is the "program bus," which controls the input to the program monitor and to the tape machines or to the broadcasting facility. Just above the program bus is the "preview bus," which controls the feed to the preview monitor, allowing both director and switcher to preset special effects before "taking" them on-line (to "take" is to put a camera or other input on-line, on the program monitor and the program feed to the tape machines, to the broadcast transmitter, etc.). Two or three other buses (sometimes more) are used to input sources for special effects (e.g., put camera 3 on the left side of a split screen by using one special-effects bus and camera 4 on the right side, by using the other bus).

If you have never seen a switcher, or a control room, in person, try to arrange for a tour of a local television station, cable operation, or

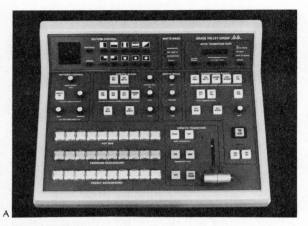

Video Switcher

(A) Although this switcher is designed for small-scale applications, it offers a wide range of features once reserved for much larger models (including downstream keying and automatic transitions by number of frames). The three buses on the lower left, along with the fader bar on the lower right, handle the cutting, dissolving, and the movement of the wipes (which are selected by push button on the upper left). The center row of buttons and knobs controls the keying, the background generators, chroma-key, and other special effects. (B) A much larger switcher, standard in most major market facilities, includes two mix/effects banks with three buses in each, four fader bars, a few dozen wipes, and the ability to combine effects in a great many combinations. *(Photos courtesy Grass Valley Group)*

independent facility. Switching consoles are extremely versatile (with improvements and new "bells and whistles" added regularly by the manufacturers), and a demonstration will introduce a wide range of creative opportunities. In fact, during the early stages of a producing or directing career, it is wise to spend as much time as possible in television studios (or with location crews, if that's your bent), observing, and, hopefully, absorbing ideas and information.

To cut from one camera to another, or one tape machine to another, the switcher need only press the appropriate button on the program bus. To dissolve from one input to another, the first input is set on one bus and the second is set on another. A lever is moved, usually by hand, to make the transition from one picture to another. Most switchers offer a variety of graphic picture transitions known as "wipes." A split screen is a horizontal wipe, stopped in midpath. Wipes are available in most geometric patterns: diamonds, squares, rectangles, circles, ovals, even hearts and stars. The dimensions of each pattern can be adjusted on most switchers, permitting an oval to look very much like a circle, or quite oblong, or a rectangle to look like either a box or a bar or anything in between. The size of each

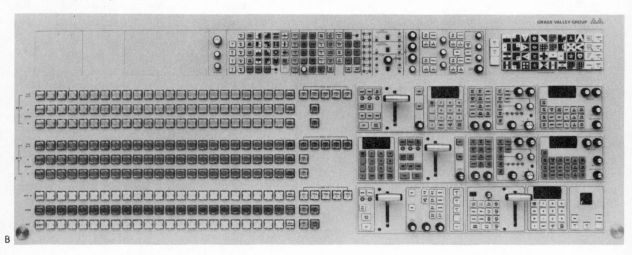

shape can be adjusted as well. A joystick is used to move the center of the wipe's design anywhere on the screen. The borders of the wipe can be left invisible; they can also be outlined and colored by turning a few knobs on the switcher. Edges and borders can be hard and well defined, or soft and hazy; they can even be made to pulsate (this pulsating business is seen on used-car commercials—it looks a little tacky). With enough buses, several wipes can be used together to create an interesting graphic design. It's usually best to preset these complicated combinations before they're needed, and then press a single button to take the wipe live (or to edit the combination in later). Since there are a limited number of buses available on any board, the same rule applies no matter how large or complex the board: "One more bus is always needed to do the wipe you have in mind."

The newest board designs include some computer technology. One feature allows the switcher to preset the length of any wipe, any dissolve, any picture transition that uses a fader bar. With digital video technology, a greater number of complicated graphic designs are available as wipes. A tour of a studio with new equipment usually includes a demonstration of some of the latest wipes and switcher effects. The latest effects are fun to use, but most of them are gimmicky and have a tendency to lose their fresh look after a few weeks of constant use by every facility with the same equipment.

Switching consoles are also used for combining elements of one picture on top of another. The most basic (and old-fashioned) of these is the "superimposition," which is accomplished by holding the fader bar halfway between one picture and another during a dissolve. In the early days of television, white letters on a black background (black tends to disappear during a dissolve) could be superimposed as a title by dissolving the graphic into a second picture. (The only problem: the letters are slightly transparent. Watch a TV show from the early or mid-1950s and you'll see.) Titles and other picture elements are now imposed by using either a "matte" or a "key." In either case (the processes are technically different, but the terms are used interchangeably), some or all of one picture is replaced by some or all of another. Using the same white letters on the black card, the white letters would now be keyed, their images replacing an identical area on the original image. The white letters would be "keyed" (or "matted") over the original, and the black would drop out. Keying can be done with colors as well, causing one or more colors to drop out. Chroma-keying is one type of keying that causes any one color to be eliminated from one picture, permitting a second picture to be used in its place. Weather maps are inserted behind weather reporters by using chroma-keys. (The process, and recent improvements on it, is explained in the next chapter.)

The operation of a control board capable of so many different functions can be complicated. Most switchers prefer to preset as many positions as possible, to save time and to avoid errors. Once the board is preset, it should be touched only by the switcher. Failure to respect this rule has cost many a director a full half-hour of resetting a complicated setup. In a union shop, a director who touches the board, even "just to see what the wipe would look like," is inviting a grievance. This is emphatically the case at the networks.

2. The Audio Engineer

The audio engineer sits behind a mixing board, usually in a soundproofed booth attached to the control room. The audio board is used to adjust the levels of all audio inputs, to select between them, and, in some cases, to alter certain frequencies. A typical board will be able to accept at least eight and as many as

Audio Booth
This audio booth looks out onto the control room, so that the engineer can see the director for cues (she hears him via intercom). Monitors permit the engineer to follow the action. The thirty-two-track audio console is more than most television productions require, but it does provide the kind of versatility that may be necessary for complicated productions that require plenty of inputs.

twenty-four different inputs, including microphones, tape machine playbacks, audio cart machines, reel-to-reel audiotape machines, and special-effects devices. Each is controlled by a "fader" (if it slides) or a "pot" (if it turns like a dial—the term is short for potentiometer), and monitored on a VU meter. (VU is short for volume units, a term that's never used. The meter measures volume, which is controlled by the pot or fader.) The level of each input must be carefully set and reset constantly throughout the production to create a balanced "mix." This mix is usually created live, and fed directly to the videotape machine or to the live broadcast facility. In concert situations, a "rough mix" is made on site (with individual inputs balanced according to the

engineer's first impressions), and the final mix is created later (with the balance on each split second of sound carefully monitored, equalized, and remixed several times if necessary until it's perfect). There's more on audio mixing in the postproduction chapter.

The audio engineer is responsible for the selection and operation of all microphones, turntables, and audiotape machines. When many microphones are required, or a boom is used, the audio engineer may employ one or more people as audio assists on the studio floor. An audio assist may physically handle the microphones (as in the case of a boom, which is moved and its microphone tilted to pick up each of the performers when he speaks). An audio assist may also operate a "sub-mixer," which is a smaller version of the audio console used in the control room. The feeds from several microphones can be mixed on the sub-mixer, feeding only one signal to the audio engineer (this makes the job easier, which may be essential on a complicated production).

Most audio engineers are versatile, trained to handle anything from a live studio audience to a marching band. With a knowledge of the wide variety of microphones and some of the more unusual ways of rigging them, most audio engineers are up for a creative challenge. In most cases, the director should leave the audio decisions to the engineer, offering occasional suggestions and making demands when necessary.

Most budgets will not permit the use of live musicians, so most music is prerecorded. The audio engineer most often uses small audio cartridges for the playback of short music pieces (under ten minutes), and prerecorded voice-overs because they're dependable and because they "cue up" (go to the beginning of the piece) automatically. Records and reel-to-reel tape machines cannot be cued as quickly (and, besides, someone can bump the tone arm on a turntable). Although the audio engineer is not responsible for videotape playback, he or she must be prepared to receive audio feeds from tape playback machines. All of these prerecorded sources are, ultimately, balanced with the microphones on the mixer.

Once again, a knowledgeable producer or director will be able to help an engineer achieve a desired effect by knowing something about sound and the way it is reproduced. When an audio engineer hears a producer or director ask for "more bass" or "more high end," or for the emphasis or de-emphasis of certain frequencies (called equalization, or EQ for short), the engineer is receiving instructions that are far more useful than "It doesn't sound right, can't you make it sound better?"

3. The Video Engineer

Although the video engineer (sometimes, the "senior video engineer") does not select the inputs, he or she is responsible for the technical quality of all of the pictures. Working with a trained eye and the help of some technical monitoring equipment, the video engineer is concerned with the proper video levels (pictures not too dark or too light), the colors (as close as possible to real life), and other tech-

Videotape Operator
On all but the smallest television productions, engineers and technicians are guided by their union contracts. All contracts set work rules, which prescribe regularly scheduled breaks throughout the day (without these rules, a director could, in theory, make unfair demands). Here, the videotape operator, located in a tape room adjacent to the control room, enjoys five minutes away from the pressure. Producers, associate producers, and others who are not guided by union contracts should learn to take breaks, but they never do.

nical specifics. Color bars (a standard test pattern that can be generated electronically) are used to test all of the colors and levels of blacks and whites. Because the video engineer is not directly involved with the director during the production (the work tends to be done alone, because the specifics are so complicated and because most video engineers are accustomed to working alone), video engineers tend to be forgotten. As a director, it is important to make a special effort to encourage the video engineer to work as an active member of the team. Spending some time with a video engineer and learning more about the science of video is a wise idea as well—everything comes in handy at one time or another.

4. The Videotape Operator

The tape operator has three main responsibilities: to record, to play back, and to keep all of the tapes in order. A good tape operator will not require much supervision; he or she will communicate directly with the tape library, will devise or help to devise a labeling scheme, and will always be available to start, stop, or play back on cue. Operating tape can be one of the more boring jobs in the studio, especially in situations where the position is filled to either play back only one or two short tapes, or when the show is simply being recorded nonstop. Once again, as a director, it is important to keep the tape operator involved, because this, too, can be a forgotten position (a situation that's worsened because most tape ops work outside the control room, usually with other operators in a central location where all of the facilities' machines are located in a dirt-free, temperature- and humidity-controlled environment).

A control room is more than a place where skilled technicians work—and is considerably more than an equipment room. It is a nerve center, where all of the necessary resources are available instantly.

5. Communications Equipment

Most control rooms are designed to be communications centers, permitting the director, staff, and crew to stay in constant touch with the studio floor, with support facilities in the building and nearby, and with the outside world. There are intercom systems, PL systems, private telephone systems, IFB systems (which are used by on-air talent at conventions, for example—sometimes, you can see the ear plug). It is important to review all of the available communications systems prior to production, and to test the most important lines during rehearsal, just to be sure.

6. The Clock

Although the clock is not usually considered to be technical equipment, it is one of the most important tools in the control room. There are two kinds of clocks used in television production: the standard twelve-hour wall clock, and the stopwatch.

Most control room clocks are tied to the facility's master clock. At the second when that clock shows, for example, 6:00 p.m., the news must begin. If the clock is a few seconds off, the news will start too early (and the audience will miss the beginning of the program) or too late (and the audience will see a blank screen). On a live show, everyone watches that clock, and countdowns are called at 15 minutes, 10 minutes, 5 minutes, 4 minutes, 3 minutes, 2 minutes, 1 minute, 45 seconds, 30 seconds, 20 seconds, 15 seconds, 10 seconds, and then down to 2 seconds (nobody actually gives a 1-second cue; the time is better spent setting up for the first cue). These time cues are fairly

Back-timing on Rundown

The producer of "P.M. America" has a problem. Due to a longer-than-planned interview with Ellen Rosenberg, and an unexpected encore for the comedy act Robert and Sue, the show is running over a minute and a half late. The second VT piece on the vice-presidential home started at 4:45:10, not 4:43:30. The show is 1:40 long, and if nothing is cut out, the show will end at 4:59:40, not 4:58:00, as it is scheduled on over a hundred stations across the country. Fortunately, the videotape is almost 8 minutes long, so there's plenty of time to think.

The first option is to dump out of the videotape early, but the segment is on to get ratings, and should not be touched. Sheila, the reporter, doesn't have to talk for 1:20 about how she felt about the house—she'll be cut back to 1:00, so we'll pick up :20. We still need 1:20.

Ken's studio audience segments are fun, and we haven't done one this week yet. If we still need some time, we can steal a little from there.

The contest plug can be cut out—we've already done it once today, and we can mention it again when we tease tomorrow's show. That's only :15, so we still need 1:05.

Most of the closing billboards can't be touched, but the ticket plug at the end of the billboards can be pulled. That's worth another :08, so we need :57. We can go with short credits, which run only :15 instead of :40, so we can cut another :25, and need only :27. And that's easy: Ken can cut about :15 out of the audience bit, and Carol can shorten the Sheila segment by cutting a question.

After double-checking the arithmetic, the producer (or director, depending on who's watching the clock) alerts everyone to the changes. They are:

1. Carol's interview with Sheila will run about :45, not 1:20 as planned. She'll cut out a question.
2. Ken's audience segment will run :10 shorter than planned.
3. No contest plug.
4. No audience ticket plug at the end of the billboards.

With a few speed-ups during the "tease tomorrow" segment, the system should work perfectly.

	Show Rundown			*Actuals*
24.	Intro Comm. #5 (Pampers)	0:05	4:40:40–4:40:45	4:42:20 – 4:42:25
25.	Comm. #5	1:00	4:40:45–4:41:45	4:42:25 – 4:43:25
26.	Intro news review (Ken)	0:05	4:41:45–4:41:50	4:43:25 – 4:43:30
28.	News review (Charlie)	1:00	4:41:50–4:42:50	4:43:30 – 4:44:30
29.	***Local news break***	0:30	4:42:50–4:43:20	4:44:30 – 4:45:00
30.	Intro VT (Carol)	0:10	4:43:20–4:43:30	4:45:00 – 4:45:10
31.	VT: Vice-president home (Pt. 2)	7:45	4:43:30–4:51:15	4:45:10 – 4:52:55
32.	Intro Sheila (Carol)	0:10	~~4:51:15–4:51:25~~ 4:52:55–4:53:05	1:40 over
33.	Sheila on VP Home (Carol)	:45 ~~1:20~~	~~4:51:25–4:52:45~~ 4:53:05–4:53:50	
34.	Studio audience (TBA) (Ken)	2:10 ~~2:30~~	~~4:52:45–4:55:15~~ 4:53:50–4:56:00	
35.	Comm. #6	0:30	~~4:55:15–4:55:45~~ 4:56:00–4:56:30	
~~36.~~	~~Contest Plug~~	~~0:15~~	~~4:55:45–4:56:00~~	
37.	Tease tomorrow (Ken & Carol)	0:30	~~4:56:00–4:56:30~~ 4:56:30–4:57:00	
38.	Closing billboards	0:50	~~4:56:30–4:57:20~~ 4:57:00–4:57:42	
39.	Credits	0:40	~~4:57:20–4:58:00~~ 4:57:42–4:57:57 (CLOSE ENOUGH!!)	

standard throughout the broadcast industry; an associate director calls the time from the control room, and the stage manager announces it on the floor.

Since broadcast schedules are arranged with several commercials and promos between shows, shows must get off the air on a precise second as well. Most half-hour shows, for example, are 28 minutes and 30 seconds; the actual running time may vary slightly at local stations.

Programs recorded "live on tape" for later broadcast are also produced by the clock. The easiest way to do this is to reset the clock for the top of the hour and treat the whole production as if it were live, making certain to finish at the right second as well. Although editing can be used to alter the time of any production, a good producer-director team should attempt to complete productions with a minimum of editing. The added pressure of live production brings a freshness to any production; editing can flatten the highs and lows and cause the production to lose some of its life.

A. Back-timing
There is a trick to making a production end on time. It is called back-timing, and it is easy to do with a little practice.

Let's say that you're exactly 20:00 into a half-hour (28:30) show, and you know that there are at least 12:00 of program material left to do on the rundown. First, subtract 20:00 (the current "running time") from 28:30 (the completed "running time"). This leaves 8:30. Second, look carefully at the 12:00 remaining on the rundown. You're trying to cut out 3:30. You can do this by either cutting a segment entirely, or cutting down the time on each of the remaining segments. With some careful arithmetic (learn to do it mentally—you'll lose time by writing it down), and constant revisions through the end of the show, it is pos-

sible to complete most programs exactly on time. Ask any news producer: back-timing is a technique that's used every night. It's so common, in fact, that many news scripts include a few optional stories of various lengths to be selected during the last commercial. (Commercials, incidentally, provide an excellent opportunity to compose your thoughts, and to tell everyone what's happening, what's being cut, and whether the show is going to be a little short [talent must stretch], a little long [talent must speed up], or right on time.)

B. Using a Stopwatch
Every director, associate director, and PA should own at least one stopwatch, with big numbers and a hand that sweeps 60 seconds (*not* 30 seconds), clear markings for minutes, and a minimum of clutter (e.g., tenths of a second, which are useless in television production). A digital stopwatch will do, provided it can be read in low light.

The stopwatch is used as a secondary clock to time videotape playbacks (if a tape is :50 long, the stopwatch starts when the videotape starts, so the talent and crew can be given a ten-second countdown at the tail), commercial breaks, and other segments.

Many control rooms have built-in stopwatches. Feel free to use them, but carry your own just in case.

DOING IT LIVE!

Before the use of videotape became standard operating procedure (in the early or mid-1960s, depending on the facility), most television programming was live. The home audience sensed that it was watching a live presentation because there was an energy, a sense of reality, a sense of being part of the show that has since been lost. When a production is recorded on videotape, and then edited, some of the magic is lost. In the worst cases, programs like "Foul-Ups, Bleeps and Bloopers" are seen as completely manufactured, and even a laugh track has a tough time making the home audience respond. In the best cases, productions like "Hill Street Blues" use the best shooting and editing techniques to tell a story that maintains its spontaneity, even though it is shot over a course of days, with scenes recorded out of order and in pieces. The feeling of most productions is not improved by editing (although "We'll fix it in the edit" is an extremely popular phrase), even though the technical and performance errors are eliminated.

A live production can have magic, but only if it is treated by the producer and seen by the audience as live. To show something live that could as easily be on tape (most live news reports on local stations fall into this "who cares?" category) is not really live television. To mount a major production, like "The Miss America Pageant" or a sporting event or a presidential debate, this can be live television at its best. When the opportunity to "do it live" presents itself to a producer or director, the pros (the magic, the spontaneity, the [slim] possibility of higher ratings) must be balanced with the cons (the possibility of mistakes appearing on the air, the chance of running long). If a live show has clear advantages, go for it! There are few things more exciting than working on live television.

The Making of "Kate & Allie"

"Working in New York isn't very different from working in California," explained "Kate & Allie" AD Sam Gary, a veteran of sitcoms like "Three's Company." "New York's not very different at all. Billy's here. New York's just colder, that's all."

"Billy" is Bill Persky, who wrote and later produced "The Dick Van Dyke Show" and created "That Girl" (both with partner Sam Denoff). Bill is an experienced professional, expert in the crafting of television comedy, and "Kate & Allie" is under his creative control. He directs every episode, sits in on story and scripting conferences, and shares the title of producer with Bob Randall (whose principal concern is the scripts).

Although the show has become his own, Bill Persky did not create "Kate & Allie." Sherry Coben, the executive story editor of the series,

© Alan Landsburg Productions, Inc.

created the series. She had written several off-Broadway plays, and for "Hot Hero Sandwich" (a clever children's series on NBC) and "Ryan's Hope" (the ABC soap), and was pitching "idea after idea" to the programming executives in New York. In the spring of 1982, she met with veteran television executive Merrill Grant regarding a sitcom concept called "Two Mommies." Grant liked the idea, and together they brought it to Mike Ogiens, a senior programming executive for CBS in New York. Ogiens liked the idea, commissioned a pilot script, encouraged her to work with a production company (Alan Landsburg Productions, a top network supplier, who had expressed a desire to produce a show in New York), and started mustering internal support at CBS.

By November, it was an on-again, off-again proposition. A second-draft script was written by Coben, working closely with Mort Lachman, who later became an executive producer of the series. There are hundreds of pilot scripts commissioned every year, but only a handful actually result in series. Not much happened until the spring of 1983, when Ogiens and Grant heard that the Creative Artists Agency, which represents Jane Curtin and Susan Saint James, was looking for a New York–based series in which their clients could both star. The "Two Mommies" script fit their requirements, the name of the show was changed to "Kate & Allie," and within a short time, CBS committed to six shows (there never was a pilot). Bill Persky was selected to produce and direct (thus providing the series with the security of a key player with a solid track record), and, thankfully, "Kate & Allie" was scheduled in a "great time slot." The first six shows were successful, and the network scheduled "Kate & Allie" for the 1984–85 season.

"Kate & Allie" is produced by a small family of performers and production personnel at the Ed Sullivan Theatre in New York City, a facility once owned by CBS, now leased by Reeves Communications (the same company that owns Alan Landsburg Productions, incidentally). The principal performers are Kate (Saint James), Allie (Curtin), Kate's daughter, Emma, Allie's daughter, Jennie, and her son, Chip. The executive producers are Merrill Grant and Mort Lachman, who attend key production meetings, work closely with the network executives, and permit Persky and the show staff to do their jobs without interference. Although some free-lance writers are used by "Kate & Allie," most of the writing, the story work, and the rewriting is handled by Bob Randall, Sherry Coben, and co-producer Alan

The cast and the director. From left to right, standing: Bill Persky, Ari Meyers, Susan Saint James; seated: Allison Smith, Jane Curtin, Frederick Koehler.

Leicht, with help and guidance from Bill Persky. Developing stories and rewriting a script until it is perfect is a full-time job, but most network sitcom writers are also contractually committed to several scripts for each season (which are written in the evenings and on weekends and between seasons, because there is no time during the production weeks). There is far more to a television production than scripts and casting, however, and coordinating producer George Barimo, a man with considerable experience both at the networks and as a series producer, takes charge of the day-to-day operation. His responsibilities begin with budgets, schedules, arrangements with the studio and editing facilities, and the network programming, "standards and practices," and administrative departments. He is also involved in casting decisions, work with art director Tom John, and with the many details that require attention on "Kate & Allie." Associate producer Cathy Cambria works closely with him. Two assistants (one for Barimo, one for Persky), two PAs, a bookkeeper, and a few gofers fill out the program's staff.

The logistics involved in producing "Kate & Allie" are fairly typical of network situation comedies. A typical sitcom costs about $300,000 to $400,000 per half-hour, and most of this money is paid to the performers, the writers, and the director (this is quite different from business video, for example, where most of the money pays for the facilities and other below-the-line costs). Roughly nine to twelve weeks are required for each "Kate & Allie" episode, from the initial assignment of the story to a writer, through first, second, and final draft, and, finally, through rehearsals, shooting, and editing. Here's the layout of a typical week at "Kate & Allie":

On Monday morning, the cast sits at a long table onstage and reads the script for the writers and producers. Since this is the first time that the written words are being said by the performers, changes are inevitable. The writers on "Kate & Allie" really

VTR DATE __Tuesday, November 20, 1984__   TITLE __"CHARLES MARRIES CLAIRE"__

p. 1 of 3

CT/SCENE	SET	CAST	PROPS	WARDROBE	NOTES
ACT I Scene 1 p. 1	LIVINGROOM	Allie,Emma,Jennie, Chip,Charles,Kate	Allie's homework;clean laundry		NOTE:clothing needed for laundry SFX:doorbell
ACT I Scene 2 p.7	LIVINGROOM	Allie,Emma,Kate, Jennie,Chip,Charles	Allie's homework including biology book		
ACT I Scene 3 p.12	KATE'S BEDRM	Kate,Allie		Kate-pyjamas	
ACT I Scene 4 p. 15	LIVINGROOM	Allie,Kate,Jennie, Chip,Charles	kitchen dressed for cooking pot roast & baby carrots; skeleton	NEW DAY	NOTE:bones on skeleton are numbered with key at base
ACT I Scene 5 p. 19	LIVINGROOM	Kate,Emma,Ted, Allie,Jennie,Chip	Jigsaw puzzle-partially finished	NEW DAY Allie,Jennie,Chip dressed very preppie	

"Kate & Allie" is an Alan Landsburg Production.

know and understand the characters, so changes are few (when the wrong writers are hired, or the script has been poorly edited, these sessions can be chaotic; at "Kate & Allie" just about everything runs smoothly—it's a pleasure to watch). Changes are made, and Persky offers some suggestions to the performers as to emphasis and timing. The results are immediately apparent; his talent, experience, and sense of humor are extraordinarily well suited for this series.

After lunch, the rehearsals begin, and, by the end of the day, the entire show is blocked. The performers know their basic positions, their basic moves, and know most of their lines by heart. Some scenes are funny as written (e.g., the lines themselves are funny), most come to life as the performers work with Persky and learn how best to play them. Although Persky moves the schedule along, the mood is easy and fun. When Jane Curtin's baby comes to visit, everything stops for twenty minutes, and it's easy to see the family feeling of the "Kate & Allie" staff and crew. Then, it's back to work, and by the end of Tuesday (which is rehearsals, all day), the show is ready to be seen by the network (Mike Ogiens and Josh Kane, the two CBS executives most responsible for the success of "Kate & Allie"). The session is loose; everybody seems to be enjoying themselves. Although there are a few rough spots (two of the kids are running around backstage and one misses a cue; Persky says, "They are *kids*, they're entitled to act like kids once in a while"), the show has taken shape in only two days. Ogiens and Kane ask a few questions, make a suggestion or

two, and show that they are pleased with the show so far.

Wednesday morning finds the cast working through one last rehearsal without cameras, polishing certain scenes, running through the show one more time. Persky and his AD (and good friend) Sam Gary spend the afternoon working out camera shots and marking their scripts. Gary takes the time to create shot sheets for each of the cameramen as well. The cast memorizes their lines; tomorrow, there will be no scripts allowed onstage.

On Thursday, the performers work in front of the cameras for the first time. Persky no longer works onstage; now he works in the control room and communicates with the cast through Maureen Thorp, the stage manager. Once again, there are problems that become evident when the lines are read and the actions are made on camera, so the day is devoted to making changes and refining the entire program. By the end of the day, the show is ready to be seen by Ogiens, Kane, and their staff of programming people. Again, there are a few suggestions, but the show is taking shape nicely. (If the show was not up to par, the problems would have started to surface when the network first approved a script weeks before.)

Friday is show day. The morning is used for a final blocking and rehearsal. Each time, the show gets a little better. Lines that felt "a little clunky" earlier in the week are brief and to the point. Scenes that seemed "long" earlier are now trim. There are no more awkward movements, no more awkward camera angles. By the time the audience arrives for the dress rehearsal, everyone is confident. The show works.

Getting a hundred-odd people to understand what's happening in a television studio—and to laugh when you want them to laugh—can be a neat trick. Normally, a comedian comes out about fifteen minutes before the show starts, talks about the show, asks how many have seen it, how many members of the audience are from out of town, and tells a few jokes. This is called the "warm-up." After J. J. (the comedian) does his warm-up, Bill Persky comes out and introduces the cast. The audience applauds as he introduces Jane, Susan, and the others (it's good for an audience to applaud—they feel a part of the show). The kids are the best he's ever worked with, because they're real kids. Then Persky routinely asks Allison Smith (Jennie) to sing a few bars from her Broadway role of *Annie,* and she BELTS " . . . the sun will come out TOMORROW." The audience goes crazy, and the show is ready to start.

"Kate & Allie" is recorded with four cameras, and although Persky cuts the show live, all four cameras are ISOed for later flexibility in editing (veteran director Jay Sandrich does "The Cosby Show" the same way). Audience reactions are recorded live, and later "sweetened" in LA ("sweetening" is the process of adding a laugh track with other audience reactions).

After the dress rehearsal, the cast huddles in the wardrobe area for notes from the director. The audience did not react to some of the lines; Persky explains that the problem was the audience, not the performances. The cast trusts him completely; he has their confidence. Some of the lines could play better, more emphasis here, don't turn so quickly there, wait for him to get up before you react, etc. A new audience (usually one that's more responsive because it is composed of people who work during the day and are visiting "Kate & Allie" as their evening's entertainment, not because some kid handed them a ticket on a street corner) is brought in for the "air show." Generally speaking, the air show closely resembles the finished product.

The week ends for the performers, who will face a new script on Monday (every fifth week is off, to recharge), but the producers, director, and staff continue because this show is not finished, and because there are more shows to get ready. The off-line editing on "Kate & Allie" takes a day or two, and the on-line takes about a day. After the show is cut, it is sweetened, either by flying the tapes to LA to a sweetening room or by flying the sweetening machine and its expert operator to New York. The network gets copies at the end of each day of editing, making changes if necessary.

"Kate & Allie" episodes are not made one at a time. There are usually five to ten episodes in various stages of development and production at any given time (fewer toward the end of the season, of course). While one script is being written, another is being reviewed and handed back to the writer for a second or third draft, another is being reviewed by the network, and the two upcoming scripts are in the hands of the performers. While this week's episode is being rehearsed in the studio, last week's is being edited, and the episode from the week before is being sweetened, or receiving final network approval. A total of twenty-two scripts were ordered by CBS for the second season of "Kate & Allie"—the team of Randall, Coben, and Leicht spend most of their time crafting each of these—to be shot over the course of about six months. And, of course, it isn't enough to just write, and rewrite, the scripts and shoot them. A program like "Kate & Allie" stays on the air because it is consistent, because it is funny. And who decides what is funny? Initially, the writers do. Then the director decides, making changes if he feels that they're necessary to get the laugh, or to establish character or situation. Finally, the performer decides what is funny, because she won't say it unless she honestly believes that it "works," even with the greatest of assurances from the writers and the director. And, ultimately, it is the audience who decides. Based on the ratings for "Kate & Allie"'s first and second seasons, the audience seems to like what they see.

KATE & ALLIE
Show #13—CHARLES MARRIES CLAIRE
VTR DATE: Tuesday, November 13, 1984

THURSDAY, NOVEMBER 8, 1984

8:30 AM–	Offline Show #11
10:30 AM– 2:00 PM	Rehearse Show #12
2:00 PM– 3:00 PM	Lunch
3:00 PM– 5:00 PM	Rehearse Show #12
5:00 PM– 6:00 PM	Run Thru Show #12
6:00 PM	Notes/Rewrites

FRIDAY, NOVEMBER 9, 1984

9:00 AM–	Offline Show #11
9:30 AM–10:00 AM	Prod. Meeting, #12 & 13
10:00 AM– 2:00 PM	Rehearsal, Show #12
2:00 PM– 3:00 PM	Lunch
3:00 PM– 6:00 PM	Rehearsal, Show #12

MONDAY, NOVEMBER 12, 1984

8:00 AM– 9:00 AM	ESU/TECH SET-UP
9:00 AM– 1:00 PM	Camera Blocking
1:00 PM– 2:00 PM	Lunch
2:00 PM– 4:30 PM	Camera Blocking
4:30 PM– 6:00 PM	Dress Rehearsal Show #12
6:00 PM–	Notes/Rewrites
	Online Show #11

TUESDAY, NOVEMBER 13, 1984

9:00 AM–10:00 AM	ESU/Tech Set Up
10:00 AM– 2:00 PM	Camera Blocking Show #12
2:00 PM– 3:00 PM	Lunch

(continued)

(continued)		
3:00 PM– 3:30 PM	Costumes/Notes/Check-In Tech	
3:15 PM– 4:00 PM	Audience Load-In/Warm-Up	
4:00 PM– 5:30 PM	Tape Dress Show #12	
5:30 PM– 6:00 PM	Audience Load-Out/Notes	
6:00 PM– 6:30 PM	Audience Load-In/Warm-Up	

6:30 PM– 8:00 PM	Tape Air Show #12
8:00 PM– 8:15 PM	Audience Load-Out
8:15 PM– 9:00 PM	Pick-Ups

THE REMOTE PRODUCTION SCHEDULE WILL FOLLOW SHORTLY

ALAN LANDSBURG PRODUCTIONS
OFFICE: 1697 Broadway,
Suite 307, NYC

!
Alan Landsburg Productions, East, 1697 Broadway
NYC, N.Y. 10019 212-307-4894
"Kate & Allie" is an Alan Landsburg Production.

SEVEN

Advanced Production

Advances in equipment design and in digital technology present the producer and director with a wide range of options. It is now possible to produce an entire program or series without entering a studio, without using real performers or props, without building scenery, without many of the "givens" that limited producers and directors earlier in television's history.

There are two areas covered in this chapter: special effects and remote production. These are admittedly different disciplines, but they share one basic truth. Both areas are more sophisticated variations on simpler production schemes. Digital effects are fascinating, but they are most effectively used by a producer or director who is well trained in a simpler form of special effects, those which require nothing more than imagination. Remote production is as complicated as it is glamorous, but a producer or director who has experience in single-camera remote work will oftentimes see possibilities during the course of a multicamera remote that those without single-camera experience might not see. In either situation, a combination of good training and good instincts will allow an up-and-coming producer or director to take advantage

of opportunities. Heading into a multicamera remote without the proper experience may be inviting disaster; developing a broad range of experience, whether gained by training or observation (if you want to direct basketball, get out there and watch a director do it a few times; you'll be surprised how much you'll learn if you watch carefully), increases the odds of success, and encourages the crew to help a novice look like a seasoned professional.

SPECIAL EFFECTS

There are three different kinds of special effects used in television: (1) mechanical effects, (2) switcher effects, and (3) digital effects. Although considerable attention has been given to digital effects (e.g., those that depend upon a computer for their magic), the first two categories offer considerable flexibility, especially on low-budget projects.

Mechanical Effects

1. Artcards
In the "early days," prior to 1963 or so, there were only two ways to show a still picture, or

a title, on a television screen: by pointing a television camera at artwork or a photograph mounted on a piece of heavy cardboard and supported by an easel, or by projecting a slide into a camera lens by using a device known as a telecine.

The card-on-easel technique was, and still is, the most useful. Rubber-cementing a sheet of paper onto a piece of cardboard is quick and easy and, besides, it's cheap. This "artcard" can be used to show the cover of a book, a news or publicity photograph, a prize on a game show, an illustration or a graphic. In most cases, a camera leaves the action and swings around to cover an easel to take the shot. There is a big benefit in flexibility here as well—the shot can be easily composed, and the shot can be changed by an on-camera move (e.g., zoom-in, pan). This is not as easily accomplished with the more sophisticated graphics setups.

The only problem with artcards is that they cannot easily be used in sequence on a live show or a live-on-tape show without some modification.

The simplest way to use two or three cards that will be seen in sequence is to arrange them side by side, center the camera on the middle one, and then pan. With more than three cards (up to nine), cards should be set in rows of three, with the fourth card mounted directly below the third, so that a tilt is the only camera move required. With more cards in a row, the images on the cards farthest from the camera will "keystone" (e.g., begin to show the angles caused by perspective).

If two cameras are available, the cards can be stacked on two easels (camera 1 would take the odd-numbered cards in the sequence, and camera 2 would take the ones with even numbers).

If only one camera is available, the topmost card on the easel can be flipped off, revealing the next card in the sequence. This is hard to do without disturbing the whole pile, moving

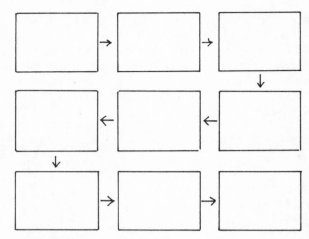

Artcards
Recommended camera coverage pattern for multiple artcards, mounted on wall or flat. Be careful of working with more than three rows or columns: keystoning will result. If there are four cards or less, use two across or two down.

the easel, accidentally showing the back of the card, and so forth, but it is possible. This technique was commonly used in the early days of television (and, in fact, many of the older stagehands and stage managers have excellent "flipping technique"). Artcards were once called flipcards; the term is still used by some art directors.

The easiest way to show a series of stills in sequence is to record all of the cards on tape, and then edit a master reel, arranging the graphics in the order that they will be played back. Although this may seem to be an expensive solution, it can be used to eliminate a camera and a crew member during the production, provided that the AD and the tape playback operator can coordinate their efforts so that the tape is advanced to the proper picture at the proper time. It's tricky, but it can work, and it does save time and money under certain circumstances.

A director can add some life to still pictures by adding camera moves. In a sequence composed primarily of stills, two cameras should be used. Each one zooms in close on an interesting area of the still, then pulls back, sometimes panning, sometimes tilting to add some visual variety. If the moves are graceful, and the music or announce track that accompanies the sequence is well prepared, the result can be quite effective. If the stills are black and white, images can be tinted by superimposing some background color from the switcher; photos taken prior to 1910 may be more effective if tinted slightly brown, for example. The addition of color is most easily accomplished by mixing a background color with the image, a switcher effect described later in this chapter.

Some producers and directors have learned to be quite clever in the use of artcards, using principles of three-dimensional design, window cutouts that show images from another card, moving parts to simulate animation (as you might find on a greeting card with a pull-tab), and chroma-key (see entry later in this chapter). When money is not available for the use of fancy digital-effects devices, producers are left on their own. A talented producer with a good director and art director can create effects that are truly magical, and, in many cases, more effective than the best that computerized devices have to offer. (Using these mechanical techniques along with the computerized gadgetry takes the concept of special effects to an even higher level.)

2. Telecine

With the development of the telecine ("tel-uh-see-nee," the more technical name for the "film chain") in the 1950s, the images from both film and slide projectors could be translated to television pictures. Slides are easy to handle and easy to store, and do not tie up a camera for graphics, so they are frequently used for graphics on local news shows. On most film chains, one slide projector is used and the between-images black is a sloppy as it is on a home slide projector. On some film chains, two slide projectors can be used to create smooth picture transitions. A slide seen from a film chain is a full-screen image; it is not possible to "start with the logo and then pull back to show the whole product" on a slide, but it is possible to do this with a television camera.

Switcher Effects

Just about every television switcher is built with some special-effects capabilities. As a rule, a switcher can fade an image in or out, dissolve from one image to another, superimpose one image on another (this is accomplished by stopping in the middle of a dissolve), and "wipe" from one image to another. Each of these transitions is made in the same fashion, by first inputting the cameras (remote feeds, tape playbacks, etc.) on mix buses A and B (A for the first image, B for the second), then selecting the mix/effects input on the program bus, and finally moving a fader bar to make the transition between the input on mix bus A and mix bus B. The speed of the transition is determined by the movement of the fader bar. (On the more sophisticated switchers, this movement can be done by computer).

1. Wipes

Most facilities have at least one full-featured switcher, capable of a great many effects and variations, with several dozen wipe patterns available. The most commonly used wipes, besides the boxes and horizontals and verticals found on all switchers, simulate the action of shutters, wedges, the hands of a clock, and so forth. Wipes are actually created on integrated circuit boards; the manufacturers of switchers generally offer a wider variety than

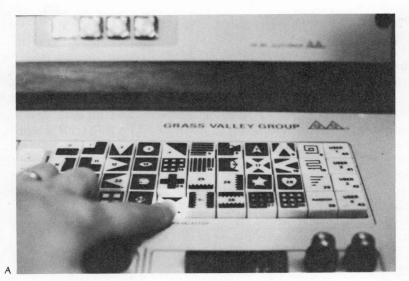

A

Wipe Patterns from Switcher
Most switchers offer a selection of wipe patterns, available to the operator by pressing a button (A). By manipulating a fader bar, it is possible to create patterns of increasing (B) or decreasing (C) size, and to develop interesting transitions (D). Patterns can be placed anywhere on the screen by adjusting a "positioner" on the switcher.

those used on any single switching console (therefore, specific wipes can be ordered by the engineering staff if they are available for the current switcher design).

Wipes can be used in two ways. First, they can make a moving transition from one image to another. Second, they can be used as frames or graphic patterns by stopping the transition midway between two images. Either way, any wipe can be made more versatile with a border; most switchers permit the user to determine the width of the border, its color, and the sharpness of its edge (soft edges are extremely effective for dream sequences; hard edges are frequently used to frame news graphics). A wipe with a border can be used as a frame or as a graphic design feature.

Switchers with border generators usually include a device known as a pattern modulator, which permits any wipe to have wavy edges that can be made to move at various speeds. This device has proven to be almost totally useless, except in the production of some sleazy commercials. (The exception may be found, for example, in the rather specialized and extremely clever use of pattern modulation in combination with a soft-edged wipe

and an aqua-blue generated color used to make a visual transition between a real ocean and a simulated one.)

Most switchers can generate any of the border colors for use as full-screen images. Using only the switcher, it is possible to fill a screen with color, alter its intensity, its hue, its saturation, and its brightness, add a border, and create a sophisticated background and frame without leaving the control room. A character generator (see entry later in this chapter) is used to print the letters in the message to complete the graphic.

A "wipe positioner" permits any wipe to be moved off-center to any position on the screen. A box wipe, repositioned to the upper-left or upper-right corner of the screen, is generally used to show news graphics just to the left or right of the anchor. A circle wipe, repositioned to the lower right, might be used to highlight a countdown clock during a sporting event.

2. Keying

Keys offer additional control over the video images. When one image is wiped onto an-

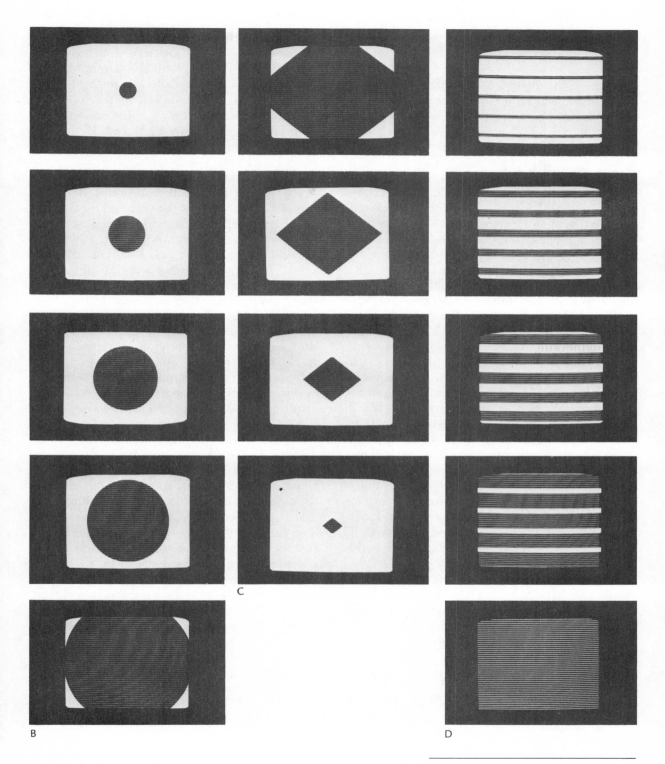

B

C

D

other, the pattern of the wipe determines the areas of the screen devoted to each image. When one image is keyed onto another, the design of one of the images determines the areas of the screen devoted to each image. In the simplest example, a black-and-white card is placed in front of camera 1. By keying, all of the blacks on the card can be replaced by images from camera 2 (or another camera, tape machine, switcher-generated background color, any video image at all). When television was black and white, a performer dressed principally in light colors could be placed in front of any background by working in front of a black drape (on camera 1), while the new background (probably an artcard on camera 2) was inserted in place of the black. Ernie Kovacs was a master of these techniques; watching his old shows provides an important lesson in the clever use of television. Kovacs was always experimenting. Decades later, his work is still fresh, still very funny. Showtime and PBS have run Kovacs best-of shows; a home video program is available as well.

Then came color. A color image on a black background could still be keyed onto another image. But color introduced a whole new set of problems. An area that would appear black on black-and-white television (reds, dark blues, dark browns, dark greens) appears in its natural colors on color television. So certain colors would key, certain colors would not key, and certain colors seemed to key well under certain circumstances, but apparently not in any predictable way.

Keying is still useful in adding titles, and in more sophisticated special-effects work on color television. But the principal use of keying—dropping out a background in favor of another—required a new technology.

3. Chroma-key

Chroma-key is a technology used on "60 Minutes," and on almost every news and sports program on the air. A single color, usually a cobalt blue or a kelly green, is used as a background color (both are highly saturated, what commonly would be called "bright blue" or "bright green"). Nothing else on the set is similar in color (no blue ties, no blue coffee mugs; no green blazers, no green plants). With a few adjustments on the switcher, the blue or green is replaced with a different image. This technique is commonly seen during weather reports, where the weather reporter stands in front of a chroma-key flat (e.g., painted either cobalt blue or kelly green) and refers to the map (either an illustration or a computer-generated graphic) behind him. On "60 Minutes," each story begins with Morley Safer, Mike Wallace, or one of the others speaking in front of a pair of magazine pages. In fact, Morley is in front of a chroma-key drape (which extends to the floor, very much like a cyc), and the magazine pages are photographed separately and chroma-keyed into place.

On a children's program, when a performer appears to be seated on a rock in the middle of gurgling brook, she may in fact be seated on a box, painted chroma-key blue or green, in front of background that's the same color (on camera 1). The brook may be nothing more than an artcard (on camera 2) with some additional switcher effects. All of the blue or green drops out of the original picture (camera 1), and is replaced by the illustration (camera 2). The net result is a real-life performer in the fantasy environment of the illustration. Using chroma-key in this fashion can be tricky, because it is impossible to know how well the effect will work before seeing it on a monitor.

Work with chroma-key backgrounds always requires expert lighting; more sophisticated uses of the technique require considerable skill and plenty of time. Poor chroma-key work is evident in the tearing between the images (this is oftentimes seen when performers have frizzy blond hair—the chroma-key blue or green sneaks in between the light strands and creates a staticlike effect). Every chroma-keyer (every keyer, in fact), has a "clipper," which is used to adjust the sharpness and the definition of the edge of the keyed area.

One further point about chroma-key, which is sometimes forgotten even by the best directors and designers: chroma-key can be used with *any* color, not just cobalt blue and kelly green. The reason why these colors are used is simple: these colors never appear in human fleshtones (using a chroma-key red can be dangerous if the performer wears red lipstick, because her lips will key out with the background; chroma-key yellow might be too close to fleshtone, and will certainly be a problem with blond hair). Alternate colors can be used when there are no people in the shot.

In recent years, there have been several improvements over the basic chroma-key process. When a performer throws a shadow on a chroma-key painted floor, a normal chroma-keyer will not show the darker areas as shadows (some chroma-keyers will ignore the darkened areas and replace the area along with the rest of the image, while others will tear at the shadow areas, demanding more even lighting). The newer Ultimatte process will show shadows as shadows, permitting a more realistic, three-dimensional composite image.

4. Multiple Reentry Switchers and Downstream Keying

When there is a need to put a great deal of visual information on the screen at once, the versatility of the switcher's design is put to the test. Even the most sophisticated switchers have only four or five buses, including the program bus, which means, in theory, that only three different effects can be combined on a single screen (the program bus must be used to "take" the effects composite, and so, as a general rule, it cannot be used to build the composite as well).

This is best illustrated by example.

The topmost bus (mix/effects A) is used to generate a background color.

The image on the second bus (mix/effects B) is brought in as a box wipe, over the background color (from bus A). The box placed in a corner of the screen is given a hard edge and a colored border (by the border generator). The image within the wipe is a videotape input; we are building a movie promo, and the clip will be seen in the window.

The third bus (technically, the preview bus, but it can be used to build effects as well) is used to add the name and the logo of the movie, which is taken from an artcard on camera 1.

We have now created a "triple reentry effect," using three buses to create the composite image. And everything is fine until the head of promotion happens to walk into the control room and asks why there is no airtime shown on the screen. The board is "tied up," there are no more available inputs, and the artcard cannot be redone in time to meet the deadline.

The solution is a device called a downstream keyer, which is a keyer designed to add titles (and almost anything else) to the image *after it has passed through the program bus*. The downstream keyer is most often used in conjunction with a character generator (see below), which types characters onto the screen. It is called a "downstream" keyer because the stream of video passes, theoretically at least, from the topmost buses, down to the program bus, and then down to master control or to a tape machine.

Triple reentry systems (and double reentry systems, their junior counterpart, also used in many facilities) are powerful tools that can be used to pile image upon image. By videotaping one set of effects and then replaying the sequence as the input for the topmost effects bank, even more effects can be added (effectively doubling the capability on the first pass on videotape, tripling it on the next, and so forth until the image on the videotape begins to degrade from multiple generations). As with chroma-keying, effects requiring multiple switcher reentries require expert knowledge of the equipment involved, and plenty of time to experiment in order to achieve the desired effect.

5. Quad Splits

The "quad split" or "four-way split" is another switcher option, available for most of the popular models. It permits four different images on the screen at one time, each from a separate bus. On most switchers, the horizontal and vertical splits meet at the center of the screen, cutting it into four evenly shaped rectangles. On some switchers, the horizontal and vertical lines can be moved, to create boxes of uneven sizes. In either case, the borders may be enhanced, softened, and colored on many switchers.

Digital Effects and Imaging Tools

Until 1970 or so, any letters or numbers shown on a television screen were first created on an artcard. Guests were identified by "lower-third IDs" (which appear centered on the bottom of the screen) that were either white-on-black artcards or slides. Credits at the end of the show (listing the name of the producer, director, technical staff, etc.) were either black-on-white artcards (with either on-camera pulls or the use of two cameras), slides, or, most often, a scroll of paper that moved upscreen via motor or mechanical device.

1. Character Generators

The character generator is a computer of remarkable versatility. In its simplest form, a character generator is a computer whose keyboard can be used to type letters and numbers onto a television screen. These characters are usually keyed over a background color or over live action (e.g., a lower-third ID). A cursor is used to position the characters anywhere on the screen.

The character generator sees each television frame as a "page," which can be numbered, stored on a computer disk and later replayed or modified by simply typing a code number. A series of pages can be seen in sequence by identifying the first page number and then tapping a key marked "next" for each subsequent page. A credit crawl can be created by another command (which is "roll," not crawl, on most popular systems). A single line of type can be made to run across the screen (as a bulletin might), its source being one or any number of pages in sequence (this is called a "crawl"). The contents of each line can be centered with the touch of a button; the entire page can be centered with the touch of another. Individual characters can be revealed one character at a time, one line at a time, or as full pages.

The ability to work with a large palette of colors is also commonplace on the latest character generators. Although most systems will permit no more than six or seven colors to be used at one time, it is possible to change colors while working on a single page, so the palette is theoretically limitless. Current computer graphics technology provides thousands of colors; most character generators offer a few dozen (which is usually enough for any job), and the latest designs offer many more. These colors can be used not only for the characters, but for backgrounds created entirely by the character generator, and for outlines that can be generated by the system as well. Drop shadows (of varying widths) are also standard, though rarely in colors.

Most character generators are now capable of using a variety of "fonts" (type styles and type sizes). Typically, a character generator is

Chyron

A character generator keyboard (A) with special keys assigned to font and color selection, centering, scrolling, crawling, and, on the right, storage and retrieval of individual "pages," as on the screen on the right (B).

sold with a standard series of fonts, which is later amended with additional designs. Helvetica is, by far, the most-used type style and it is available in the greatest number of sizes (including very small and very large; most fonts are available only in medium sizes). Each font includes both standard and italic type styles. Most manufacturers will create special fonts on demand (for a fee), permitting broadcast stations to call up a variety of logos as individual characters (the letter *A* on the keyboard might be assigned to "WNBC," the letter *B* to an NBC peacock, the letter *C* to a small-sized [channel] 4, the letter *D* to a larger-sized 4, and so on). Font creation can be done by the facility as well (it requires only a camera, a character generator, and some relatively inexpensive software), but the procedure is tricky and can be extremely time-consuming (it is the sort of work that a friendly engineer might come in and do on a weekend, when the facility is relatively quiet and, if the producer is lucky, time is not charged by the hour).

The most popular character generator is manufactured by a company called Chyron (KY-ron), whose name is sometimes used as a synonym for "character generator."

2. Digital Frame Storage

These "electronic slide projectors" have been around since the mid-1970s, capable of storing thousands of single-frame television images on computer, and then offering each one for playback by number. A graphic is easily recorded by pointing a television camera at an artcard and then storing the image by pressing a "record" button (thus eliminating the "flip-card" problems discussed earlier in the chapter). The graphic can also be enhanced with switcher effects, or with digital video effects (described later in this chapter). The frame storage device makes no distinctions between the source of its video; it simply records what it sees on-line when the record button is pressed.

This still-frame procedure is digital; the picture information is stored as a string of data bits (in much the same way that a word processor stores words and sentences as bits, but considerably more memory space is required).

The "ADDA" is one of several digital frame storage devices on the market. The term "ESS" may also be used; it is short for the generic "electronic still storage."

3. Quantel and Other Digital Video Effects Devices

Shortly after the digital still-frame technology was developed, devices that could alter the size, shape, and orientation of moving video images were introduced (these devices see video as thirty distinct frames every second, but work quickly enough to appear that they are operating on moving pictures). Generically called DVEs (for digital video effects), these magical products have appeared under the names Squeeze Zoom, Quantel, Mirage, and ADO. Each requires a large computer and sophisticated software to manipulate images in some very unusual ways.

Once an image is digitized, it can be stretched or squeezed, reduced in size, enlarged (with some loss of clarity, as in a photographic enlargement), and moved to any screen position. These capabilities, common to most DVEs, can be seen on commercials, as transitions on entertainment programs, and, with some panache, as an important part of the presentation on the syndicated "Entertainment Tonight." The most useful combination of these functions permits a director to reduce a full-screen graphic (or videotape) to a small-sized box that can be inserted (usually in a box wipe) beside any newscaster. Just about every news program uses this technique.

The early DVEs could do everything described above and only a little more. The most useful of the early enhancements is a perspective/axis system that makes the image appear to be angled away from the viewer. The degree of the angle can be controlled by number. And, on a Quantel with two channels (one of several available enhancements), two pictures can begin to create the effect of a tunnel. When one image seems to move 90 degrees to reveal another, one of several digital devices is creating the effect (this move is available with increasing variety and sophistication on the newer devices).

A more advanced version of the Quantel permits what video artist and design engineer Dean Winkler calls "two-and-a-half dimensional image manipulation."

The perspective control is more sophisticated on the Quantel 5000, which permits images to appear not only on horizontal and vertical axes, but on diagonals as well. Software packages are now available that will allow this kind of device to fly images around the screen.

Quantel's Mirage is even more sophisticated. With it, television pictures can be wrapped onto a sphere, a cone, a cylinder, any shape that can be described in three-dimensional geometry (an x, y, and z axis, if you recall your high school math).

The ADO, for Ampex digital opticals, is one step beyond the Mirage. It adds shadows, creating a very eerie sense that the three-dimensional effect is not an effect, but something real. The ADO also permits the placing of images next to one another, so that a cube might appear to be composed of six different television images as it spins (another ADO capability) in three-dimensional space.

Digital effects technology is moving quickly, with new devices and major enhancements introduced every year. These devices are expensive (several hundred thousand dollars; roughly equivalent to the cost of three or four new 1-inch tape machines, four or five ENG cameras, or a fully equipped mini remote van). Most facilities buy one, and use it for as long as possible (competitive forces, not obsolescence, are the usual reasons for updating or replacement). The latest devices are usually found at the largest facilities in the major markets, particularly in the network sports departments and in the independent postproduction houses.

One further word of caution about the use of digital effects devices (and all special effects). Many of these effects become quite popular when they are first available (e.g., flipping pictures over on a Quantel), and then become trite. As a producer or director, it is usually best to use an effect early in its life-span and to discard it before everyone else catches on. The fresh look of a production may be destroyed by the inappropriate use, the overuse, or the use of dated special effects. Most are useful as attention-grabbers, in openings and in promos, and only rarely fulfill a specific production need (e.g., reducing the size of a news graphic so that it may appear beside the anchorman). Using special effects in the body of a show, except in special situations, can be distracting.

4. Computer Graphics Systems

Digital effects devices are, essentially, image manipulators. The process begins with an existing image, which is digitized and then rearranged by a combination of hardware and software.

In 1983, several "paint box" systems were introduced. Each system permitted, for the very first time, the creation of images without a television camera, without a videotape machine, without any traditional television devices. A paint box is a computer graphics system that includes a powerful minicomputer, a graphics tablet and stylus, special software, and a high-resolution color monitor. The paint box is used to create single frames of video in much the same way as a graphic artist creates images on a drawing table. The difference, of course, is in the speed and the versatility of the computerized system.

Anything that can be done on an artist's table can be done with a paint box system. The stylus can be used as if it were a paintbrush (with a variety of brush widths), an airbrush (with an adjustable nozzle and no need for cleanup), a stick of chalk, a blade (to cut out a portion of the image and paste it elsewhere). It can be used to cut stencils, to outline, to draw ruling lines, to create boxes and rectangles and ovals and almost any other shape. If the stylus is being used for drawing or painting, its tip can pick up any one of several hundred thousand colors, which can be taken from any television picture (if you like the color of someone's gold ring and want to pave the street with gold, simply touch the stylus to the gold ring, with one simple command and start painting). Colors can be mixed on an on-screen palette, made lighter by adding some white, darker by adding black or another dark tone, shaded with any color at all. Any portion of the image can be enlarged or reduced, and repeated anywhere on the screen. Image A can overlap B, or B can overlap A with a single command. All of this can be done quickly, and easily, with a minimum of instruction (the commands are all on menus—just touch the stylus to the menu item to make a selection).

Although the paint box can be used to create its own images, it is often used to enhance or alter existing pictures. A single frame of video becomes a paint box graphic (which can then be airbrushed, stenciled, or otherwise treated to the paint box's magic) by a relatively straightforward technical procedure. The single frame of video can come from any video input—camera, videotape, or switcher. The combination of real-life images and a high-power computer graphics system can be used by a talented artist to create some wonderful, and unique, images. And unlike the other digital video devices, the paint box is completely "transparent." Because the device takes on the style of the artist at the graphics tablet, the paint box has no look of its own. There are no effects to become trite and tired from overuse. The paint box is an enormously useful tool, more closely related to the tools normally found in an artist's studio than in a video control room.

The only problem with the paint box is its inability to work with more than one frame at a time. At present, each video frame must be created (or enhanced) individually. With future paint box systems, the computer will remember all that remains the same from one image to the next, and the artist will change only the new material. The process, similar to film animation, will permit a paint box artist to retouch any video sequence, to add or subtract

A

B

Paint Box
(A) The artist works at a graphics tablet, applying pressure via metal stylus. Menu selections and work-in-progress appear on the screen. A computer keyboard is useful for titles and for storage and retrieval of finished pages. (B) Paint box systems allow skilled artists to create imaginative graphics. This "Entertainment Tonight" title slide provides an excellent example of work that can be done on a paint box system. *(Photo A by H. Blumenthal; Photo B copyright © 1984 Paramount Pictures Corporation. All rights reserved)*

characters, set pieces, and props, and, with some practice and sophistication, create frame-by-frame video animation.

More immediate changes in paint box technology will permit widespread use of three-dimensional imaging. In the latest models of the paint box, a three-dimensional figure can be defined, and then moved to any position in space (actually, beyond the borders of the screen). The figure can be lit from any angle or angles, shaded to the desired effect. Finally, the figure can be viewed from any (theoretical) camera position. This technology, which clearly combines the best of the digital effects devices with paint box technology, is available right now. It's expensive, and sophisticated, and it can be found only in the largest facilities, but it does exist and it is likely to be modified for more widespread use in the near future.

Other Effects Devices

There are two ways to manipulate the speed of the action: by altering the playback speed

of a videotape, or by using a special system, such as a slow-motion ("slo-mo") disc.

1. One-Inch Videotape Machines
There are several effects available on most 1-inch tape machines. When the smaller 1-inch format replaced the older, more cumbersome 2-inch format, the new machines were designed to start and stop with great precision. Taken one step further, the technology permits an accurate freeze-frame, one that can be identified by frame number (there are thirty frames per second on television, so a frame is equal to one-thirtieth of a second).

The freeze-frame is only part of the story. Most 1-inch machines can advance the action one frame at a time, at almost any speed. This freeze-frame and frame advance feature can, theoretically, be useful as a frame storage system (wear and tear on the machines caused by freeze-frames for long periods make other systems more cost-effective, however). A still frame can be useful, however, to start a news report, or to create additional impact at the end of a story.

With some accessory equipment, most 1-inch machines can play in slow motion as well. In many installations, 1-inch machines are replacing slo-mo discs (see below) for instant replays.

2. Slo-Mo Discs

Used for many years, the slow-motion machine employs an erasable videodisc in place of tape. The disc can record up to sixty seconds of video, and can access any sequence on the disc within five seconds. Once the spot is found (e.g., the beginning of the sequence to be replayed in slow motion, as in an "instant replay"), the speed of the playback is completely variable.

The slow-motion disc machine requires its own operator because every important play must be recorded for possible replay (the system is used almost exclusively on sports presentations). Although the 1-inch tape machine is capable of playing back in slow motion, the disc system is specifically designed to absorb the abuse of constant shuttling rerecording (this wear and tear can harm an expensive 1-inch machine). Until a digital device comes along (there are several in development), instant replays are likely to be handled as they have been for years, by the slow-motion disc machine and its always-on-top-of-things operator.

REMOTE PRODUCTION

Producing a program on location with multiple cameras requires considerable skill, experience, and a thorough familiarity with the dynamics of live events. Working on location at a football, basketball, or baseball game, or at a concert, parade, or political rally, demands a combination of standard studio and location techniques, an expertise in logistics, and an ability to give the audience a sense of "being there." Multiple-camera remote production demands the highest level of organization and communication, because so many aspects of the event are beyond the producer's or director's control.

There are similarities between large-scale remote production and single-camera location work, but the complexity of many large-scale remotes more closely resembles a big studio shoot produced with only a few of the controls that a studio typically provides, like soundproofing, access only to authorized individuals, and protection from the weather. Difficulties in any of these areas can be disastrous on a complex remote production.

Preproduction

It's only logical to start work by becoming familiar with the event and by studying any previous coverage. This is a relatively simple task for most sporting events, since regular schedules and an established tradition of camera placement and styles of coverage provide ample opportunity for study. If a musical act is on tour, it is wise to travel with the act and observe two or three concerts, making a "scratch tape" (e.g., a reference tape using a home video camera and recorder) for later study. If the event is a one-time affair, with no previous history of television coverage and no tapes to provide background, extensive meetings are the best (and usually the only) way to prepare for television coverage.

After the producer and director understand the way that the event is presented, they set about developing a good working relationship with the key players and their representatives. Coverage of sporting events, for example, usually involves a close association between the venue (in this case, the "venue" is a stadium or arena, but the word is also used to describe a theater, auditorium, or anyplace where an event is presented). In these situations, the location manager may represent both the team and the stadium, and one meeting with one person may be sufficient to set up the basic operating plan. Coverage of rock concerts, and conventions, on the other hand, involves the rental of a venue by a promoter. In these situations, there must be two sets of meetings, one with the promoter and another with the manager of the location. Coordination with both parties can be tricky, because the introduction of television coverage can make demands upon the venue that may not have been considered when the event was originally booked. Television coverage frequently comes about as an afterthought, so the producer and the promoter may have to work closely together to make changes in the original plan.

A series of meetings with the promoter and the venue management will provide the producer and director with enough information to start planning the production. In most cases, preproduction begins with a location survey and then proceeds with a large-scale production meeting.

1. The Location Survey

Four people are essential for the location survey: the producer, the director, the engineer in charge, and a knowledgeable representative from the location or the event. One person should bring along an instant camera to shoot pictures for later reference and for possible use in storyboarding. If the event is to be staged outdoors, the survey should be scheduled for the same day of the week, at the time of day when the event will occur. There are two reasons for this: (1) the position of the sun will tell the director where cameras can and cannot be placed; (2) patterns of traffic and noise frequently run on some sort of regular schedule: the kids get out of school and the street is crowded with school buses; most of

the arrivals at a local airport come in during the late afternoon; one of the tunnel entrances becomes an exit for the morning commute, making car horns and traffic noise a factor. A survey checklist is likely to be extensive, and should include all of the following, roughly in this order:

A. **The Basic Setup Plan.** The producer and director should decide, if the situation is flexible, where the stage, the audience, and the mobile van will be. The decision will be based on the physical limitations of the location, like topography and the ease of access from paved roadways; upon the direction of the sun, because neither the cameras nor the performers can face directly into the sun; upon security, so that the stage, the performers' private areas (dressing rooms, for example), and the mobile van cannot by reached by members of the audience (it's even better if the vans cannot be seen at all). Dressing room and production office space should be established, along with traffic patterns (e.g., from stage to mobile van, from stage to dressing room).

Most venues will provide electrical, structural, and other plans on request. Ask for these early on, because some searching may be required to find the original construction plans for the building.

B. **Permits and Clearances.** Outdoor locations usually involve mobile vans, cars, trucks, and trailers that may obstruct street or sidewalk areas for extended periods of time. Almost all municipalities have policies to accommodate television and motion picture crews; there are forms to sign, standard fees for permits. Applying for permits early is essential, because red tape can cause some serious delays. Some producers take the chance and shoot without permits on single-camera ENG shoots, managing to get the shots before the cops arrive; arranging a major multicamera shoot without arranging for the proper permits is foolish and irresponsible.

C. Location Fees. A television crew cannot shoot or set up its equipment or personnel on private property unless prior arrangements are made. The complexity of these arrangements varies with each location and each situation. Tapping into a nearby store for power, for example, might carry one fee, while using somebody's home for a production office might carry another. How much? Location fees are always negotiable, but a location that is frequently rented is likely to have some established fee structure.

D. Power. Television equipment requires a lot of power. Many venues are now built with extra circuits especially for television productions. When it is inconvenient or impossible to tap into existing power lines, a generator truck is used (these can be rented in most major cities). Generators come in various sizes, from portable (small enough to fit into a small van, with enough power to run a few VCRs) to full sized (large enough to power a full-scale mobile truck and all of its equipment, including the air-conditioning system).

Power requirements can be complicated, and are usually left to the engineer in charge.

E. Telephone Lines and Transmission Facilities. If the program is to be broadcast or cablecast live, arrangements must be made for the rental of telephone lines, microwave or miniwave links. Some of the larger facilities are equipped with transmission equipment, but most situations require arrangement or rental by the television production company (how it's done is covered later in the chapter).

F. Security. The producer of a remote production must consider the possibility of theft or other disturbance. Security must be considered in three key areas: the stage, the mobile van, and the backstage area. The key problem is keeping unauthorized personnel away from certain areas; limiting access with a badge system usually works very well, with a security guard posted at key locations. Most audience members will respect areas that are roped off, etc., but some events, like rock concerts and end-of-the-season sporting events, require extra attention to security.

Coordination with the venue's internal security system is imperative.

G. Union Considerations. Just as television productions are staffed by union and nonunion personnel, live events are staffed by their own combination of union and nonunion people. During the initial survey, find out about the venue's union contracts. If the venue has had experience in resolving potential conflicts (e.g., setting up an interface between two lighting and staging crews), the venue's location manager may be able to offer some advice. More complex union situations are, however, best handled by the producer or experienced production manager, and by the event's senior management. The producer must understand what his or her people can do, cannot do, and can do only with the permission or supervision of a union member from the venue. The producer and crew are guests in someone else's house; becoming "too comfortable" almost always leads to problems.

H. Insurance. Every venue must carry liability insurance, which protects both the promoter and the venue if someone is injured. Television productions also carry insurance, which protects the producers or the station or the facility against liabilities, protects the equipment in case of damage or theft, and generally guards against losses. Although there is a natural temptation to make assumptions about insurance coverage (it is not one of the more glamorous aspects of television production, and it is easily forgotten), it is wise to double-check policies and to coordinate coverage with the venue and with the promoter, and with the station or television facility.

I. Schedules. Two basic schedules should be set with the manager of the location: a weekly (or daily) schedule leading up to the event, and a shooting schedule that permits time for the setup, the shoot, and the strike. Schedules should be distributed to all personnel at the venue, at the promoter's office, and, of course, to the staff, the crew, and the performers. Updates must be distributed to everyone on the list.

J. Future Access. If there are to be more visits to the location, arrangements should be made for future access (especially if the location is normally closed during working hours). Now is the time to trade home phone numbers as well.

After the survey, the producer, director, and engineer discuss the specifics relating to television coverage. About a week before the event, all of their decisions are final, and a large organizational meeting, often involving the entire production staff and crew as well as representatives from the promoter and the venue, is used by the producer and the director to outline all of the plans in detail. Information packets, including schedules, floor plans, home phone numbers, and other vital information is distributed at this "production meeting," "prepro meeting," or "tech meeting" (any of the three names are common, depending upon local tradition).

2. The Mobile Van

Although the mobile van goes by many names (remote van, location van, truck, football truck), the concept is always the same: it is a control room on wheels. The mobile van is divided into three principal areas, very much like a control room, but designed with a reason for every inch of space. The audio control area is soundproofed, with space enough for one person, a console, some playback equipment, and a monitor or two. The video control area allows space for a video engineer to adjust the pictures from all cameras; in some trucks, the video and the videotape areas are one and the same. The director's area is tight as well, with enough room for the TD, a switcher (sometimes scaled down for space), a bank of monitors, and a minimal amount of desk space for production personnel. There are storage bins at the base of the truck, accessible from the exterior, where cameras and some lighting gear are stowed.

Mobile vans are typically cramped, chilly (the equipment requires air conditioning, a relief in summer, not much fun in winter), and dimly lit (with small spotlights illuminating the working areas, so that the monitors can be seen in relative darkness). Access to the inside of the truck is limited to essential technical staff, the director, an AD (sometimes eliminated due to lack of space in a small truck), and the producer (who usually stays in the truck only during the show, and runs back and forth between the stage or field and the truck in between times).

Local television stations with big sports contracts usually own at least one mobile van (in some cases, these have been replaced by a control room built into the stadium or arena, especially in newer structures and in venues where coverage is done by the same local station season after season). Fully equipped vans are available on daily rental in most large cities; vans can, of course, be driven from one city to another, as is often the case during football season, when most of the country's vans are booked all season long.

3. Working with Special Equipment

With the big-money status bestowed on network sports in the 1970s and 1980s, there has been considerable innovation in equipment created especially for large-scale remote productions. Special microphones with extraordinary sensitivity (to hear the action of a particular player on the field), cameras with superlong lenses (to see the player in close-up), clever camera mounts (like skycam, which allows the director to work beyond the bounds of fixed camera positions), miniwave transmitters (used on a "wireless camera," similar in concept to a wireless microphone), and a miniaturization of items once too large to be portable (camcorders, like Betacam, for example) are all becoming important tools for remote production.

Some devices designed specifically for field production are not new at all. A combination headset-microphone, for example, is commonly used for sports coverage. The headset is a two-eared model, with the sound of the program fed to one ear and the director's instructions fed to the other. The phones are muffed to eliminate crowd noise. A microphone is attached directly to the headset (as on a telephone operator's headset), permitting the sportscaster to move in any direction without restraint.

4. Selecting Camera Positions

Camera positions must be fixed before the crowd arrives. Once the cameras are up on their platforms, or spotted at ground level, their cables run under the grandstands, or around the field's perimeter, they are not easily moved. Camera positions must be selected carefully and, if possible, tested prior to the event.

The first position is most often a cover shot (also called a master shot), which will provide a good general view of the event. It usually follows either the ball (in a sporting event) or a key performer. The second position usually is selected to see specific action more closely: at a parade, for example, this camera would be raised only slightly above the crowd, probably near the reviewing stand; at a concert, it would be the camera that takes close-ups of the lead singer. The third position, and all subsequent positions, depends entirely upon the type of event and its coverage requirements. These extra positions should be selected to provide the viewer with a close-up view of particular aspects of the event, like the goal in a hockey or basketball game, close-ups of the pitcher, base runner, or the batter at a baseball game, shots of the backup band at a rock concert, and so forth. A hand-held camera is useful for weaving in and out of a parade, for moving around onstage, for following a particular player in a football or soccer game, providing some unusual angles and some additional close-ups that add to the sense of "being there."

Camera positions may be set at ground level, slightly raised (usually a preferred situation), or high above the action (best for cover shots). All cameras should face the same basic direction; cameras should be positioned in an arc, never facing each other because the same action covered from both sides of the field confuses the viewer (except at a baseball game, where the center-field camera, with its reverse angle on the battery, is directly across the field from the cover camera above home plate). Cameras should have lenses with large zoom ratios so that the director can choose wide-angle, standard, and telephoto shots from each camera position. Since the location survey is usually made in an empty arena, it is important to consider each camera's line of sight once a crowd fills the bleachers (and, most likely, stands up during exciting plays—an especially important consideration if a camera is at ground level).

Camera platforms must be sturdy enough to support both camera and operator. Cameras can be raised by constructing platforms, finding positions in the bleachers (or in the press box, if one exists), or by building a basket for camera and operator that hangs from the side of a balcony (this is common in permanent setups like sporting arenas). It is frequently possible to use the top of the mobile van as a camera position as well.

On an outdoor event, the sun will provide most of the illumination. During the survey, the director should notice the sun and its position in the sky, taking note of glare and shadows. Since television cameras have limited ability to show both extremely bright and extremely shadowy areas in the same picture (the bright area will bloom, or the dark area will go black—television cameras can see only a limited range of contrast), cameras must be placed so that their images can be properly balanced by the video engineer. Camera positions also must be selected to avoid any camera shooting directly into the sun. For these reasons, it is best to survey the location on a sunny day, preferably at the hour when the event is scheduled to take place. In some situations, television lighting equipment may be needed to supplement the stage or arena lights. Coordination with the house lighting personnel is important for reasons of both aesthetics and power requirements.

5. Communications

In a studio, communication is accomplished in two ways: through headset intercom systems and through the public address system. When the location is a big one (a concert at a racetrack, for example, where the performers are in the infield and the truck is parked beside the grandstand, near the power), communication can be a real challenge. The first thing to do is to hire a few runners, whose sole function is to find people and deliver messages. The second is to buy or rent a walkie-talkie, and to leave one unit on in the truck and the other on the stage during setup and rehearsal. If the event is to be staged over a large area, like a convention center or a park, and telephones are available, a wireless telephone pager can be useful as well. Once the cameras are connected, their intercoms will allow communication between camera operator and control room (the connection is made by the camera cable), but someone, usually a stage manager or a production assistant, must wear the headset at all times. A public address system on location is useful, provided that the area can be covered with one of reasonable size. If the location has a built-in PA system, it may be possible to use it during setup, during rehearsals, and sometimes during the actual production (although minor modifications may be required).

In a studio, the communications system comes as part of the package. On location, communications is one of several production elements that must be created from scratch. Checklists, and experience, fill in the details.

Access to outside telephone lines is very important; last-minute changes are common on location shoots. A good PA, a local Yellow Pages, and a telephone can solve a multitude of problems. Without telephone access, the production can become isolated and the producer can become trapped, losing valuable staff members to errands that would normally be handled by phone.

6. Transmission Equipment

There are three ways to transmit television signals from the site to the broadcast facility:

via telephone lines, via microwave links, and via satellite uplink. The first is the most common, and it requires the rental of broadcast lines from the telephone company. Most regional phone companies have a special department for broadcasters, and can make specific equipment recommendations for both the transmitter and the receiver. The second, because of the popularity of ENG units, is a relatively simple solution for most local television stations. The crew simply drives an ENG van, with its goldenrod microwave antenna, near the mobile van, runs a cable, and establishes a link with a microwave dish (which requires an unobstructed straight-line, but can be relayed from one microwave system to the next). The third system, working via satellite, requires either a link to an existing earth station ("uplink"), or, on a big-budget project, a portable satellite transmitting station, and, of course, the necessary receiving equipment at the other end. This situation is used for special situations, like the live broadcast of a major event, and should become more popular as the price of equipment goes down.

7. Scheduling

One of the toughest aspects of planning a multiple-camera remote is developing a reasonable setup, rehearsal, and shooting schedule. Under the controlled circumstances of a studio shoot, estimating the amount of time required for setup, lighting, ESU, camera rehearsal, dress rehearsal, audience coordination, shooting, and tearing down is comparatively easy to do with some experience. Scheduling for a large-scale event, like a parade or a concert, is more difficult for two reasons: first, because of the sheer number of variables outside of the production's control (weather is at the top of the list), and, second, because of the fixed time of the event, the entire schedule must be "back-timed" (regard-

less of whether the television crew is ready or not, the crowd—and the network and its affiliates—expect "The Macy's Thanksgiving Day Parade" to begin on time). An experienced producer reduces this pressure by allowing enough time for every eventuality, and by insisting upon a budget that permits the crew, the talent, and the staff to be on-hand for the extra hours that such a schedule demands.

At least one day of preset is necessary for most remotes; this allows sufficient time for the setup of any scenery or set pieces, the lights, and other television equipment. The day before the shoot should see the completion of the setup: all set pieces should be in place and working, all equipment should be cabled, tested, and ready to go with only a short ESU on the next morning, and performers should be familiar with the basic staging plans prior to the day of the shoot. On the other hand, it's best to avoid too much rehearsal; many performers thrive on the audience reaction at a live event and seem somewhat "dead" during rehearsals. A run-through is usually a good idea, if time permits.

Since the realities of television production usually demand that a great deal of work be done under serious budgeting and scheduling restraints, setups that require work through the night and into the early morning hours are not uncommon. It is the job of the producer to pace his people, to set priorities, to discuss reasonable compromises so that key people are not starting a shoot day with only a few hours' sleep. When individuals start to show that they're tired, when they get punchy, or start to slow down, it's time to stop for the night. An experienced producer or director will recognize this, and call a "wrap."

Incidentally, anything that must be left on location overnight should be entrusted to a security guard hired by the production—the stadium's or auditorium's guards should not be responsible for the television equipment.

Production

Regardless of the time of the event, the day begins early. Last-minute details find engineers and some of the members of the stage crew on site before breakfast. Arrival before call is normal, almost routine. It is common to say, "I should be there, just in case they need me."

The schedule is posted, and strictly enforced. No time can be lost to late arrivals, to sloppy work. There are always problems: the wind knocks over a microphone and none of the replacements has the right pickup pattern; one of the hand-held cameras falls off a chair and cannot be replaced in time, so four key scenes have to be reblocked and one of the camera platforms has to be moved before the crowds come in; the associate producer's purse is stolen (and inside are the keys to the hotel room, where all the prizes are being stored until showtime). Still, the show must go on, sometimes with modifications.

The day begins with an early morning technical check, with production assistants and associate producers running through checklists, disseminating information about last-minute changes. The producer runs through the routine one last time with the performers. The director marks the script one last time.

Then the clock-watching game begins. There are only a limited number of hours to take care of everything: to pretape the act that must leave before the show actually begins, to run through three tricky scenes one last time, to reblock a scene that looked sloppy during yesterday's rehearsal. Once the crowd starts to gather, the noise and the milling of people make anything more than camera registration and last-minute lighting or microphone checks nearly impossible. Moving from one side of the arena to the other is no longer a beeline; now, everyone uses the aisles. As the seats start to fill up, the excitement builds. About fifteen minutes before showtime, the staff, the crew, the performers are ready, in position to start. The AD calls time cues into his headset; there is a stage manager positioned beside the hosts' camera, but many directions from the control room are communicated directly to the hosts via IFBs (supersmall earplugs, wired directly to someone in the control room, usually an AD or AP). These final moments are used to get a sense of the event, an intangible that will be translated by the director so that viewers will feel as though they are a part of it (this is usually done with some ad-lib narration, a sound mix that emphasizes "ambient sound," and with plenty of crowd shots—especially close-ups of fans, children, crazies in the stands).

1. The Live Remote

The actual shoot, if it is live, is paced by a combination of the script and the play of the live event. Each camera position is assigned to a specific range of shots, and the director quickly finds a routine during the early stages of the coverage, making modifications, like using camera 3 instead of camera 2 for the long shot, making certain to include some crowd in the shot of the host, always working closely with the producer. During a well-organized remote, the producer spends his or her time in the truck, maintaining an overview, advising the director on shots and on the overall presentation, making decisions that are carried out by associate producers or other support staff, never becoming too involved in the minor emergencies (e.g., "putting out fires"). A producer who wanders in and out of the truck is not paying proper attention to the finished product, not providing the director with the level of support and guidance that is essential to the success of the project. He or she risks being in transit when a crucial decision must be made, and, more often than not, causes distractions by meandering in and out of the control room. One or more PAs or runners should handle the back-and-forth whenever possible. The producer must be available for decision-making.

2. The Tape Remote

The situation is quite different on a taped project, of course, because the producer must be concerned with the proper execution of individual pieces for later editing. Here, the process is a stop-and-start situation, where the producer is most useful in the truck during the actual taping, and may be essential onstage in between times.

The director should not leave the truck during the shoot, except during breaks. For directors who are accustomed to bouncing back and forth between the control room and the studio floor, this can be a source of frustration. A director working on location must learn to depend upon the stage manager and upon the technical director; no one person can control every aspect of a big event. The stage manager must remember that the director can see only what the cameras see, so he or she is expected to make suggestions at the appropriate times (during a long sequence when the director is holding on a shot, for example). A good technical director not only switches and insures the overall technical quality of the production, but watches the monitors and makes suggestions, again at the appropriate times. With a little practice, a good director-TD team will find those shots that make a good show into a great one, will develop the effects and transitions that make the presentation special. The TD can and should cover for the director at times, occasionally talking camerapeople into shots when a more pressing situation demands the director's momentary attention.

3. Working with an Expert

When an event is complicated, some directors choose to work closely with an expert, who helps to plan the coverage, and, in some cases, sits beside the director in the control room or truck. The practice is common in the coverage of classical music, where the expert, typically a conductor or arranger, readies the director for specific musical passages ("Here comes the oboe sequence . . ."), so that the director is sure to get the shot. An expert can also be helpful when an unplanned encore leaves the director, and the camerapeople, without the benefit of a rehearsal. The expert factor is also common on sporting events, particularly with directors who may not be knowledgeable about the fine points of the game. A director who is suddenly asked to sit in for the usual director of a sporting event may benefit from the services of an expert as well.

4. When a Mistake Is Made

Most remote productions are not perfect; mistakes are made because of the pressure, because voices on the headsets are not always heard clearly above the din of the crowd, because a technician (or the director) becomes engrossed in an important play and simply forgets a cue. As the person in charge, the producer or director must maintain a balance, ready to make sharp, certain criticisms when necessary, ready to hold back when the person knows that he or she has made a mistake and must now concentrate on not making another one. Working in close quarters with a crew, under pressure, requires a special attitude and a unique definition of teamwork. This is most often cultivated over time.

5. At the End of Show Day

When the day's shooting is over, nobody leaves except the audience. Striking a location is time-consuming, and most crews will happily accept the help of producers, directors, and their staffs in packing up for the night (this is the one time when union regulations are

Don Carney Covers the New York Yankees

"I was there when it happened," explains Don Carney. "I was working at WPIX as an AD in 1953, the station was emphasizing sports, and when they started covering the Yankees, I was the logical choice for director." Carney is now both executive producer and director of the Yankees telecasts, one of the best in the business.

He came from vaudeville, was studying music at a college in Boston, and naturally drifted toward radio, first as a performer and then as a director. Shortly after he arrived in New York, a friend asked him to fill in as a television director. Never having directed television, but clear in his understanding that the director is expected to lead, he entered the studio, looked at the crew, deadpan, and said, "I don't want to change anything—just do it the way you normally do." His career in television had begun.

Today, Carney covers over one hundred Yankee games each year (during the winter, he does football and other sports for the station). Since WPIX is a superstation (it is carried by a few hundred cable systems nationwide), millions of people see every game. And the games are colorful, because Carney sees baseball coverage as "entertainment," and because of the colorful manner of the outspoken Phil Rizzuto, a former Yankee shortstop who has a reputation for speaking his mind.

On home games, Carney works from a control room in the basement of Yankee Stadium. Away games are covered in a rented truck.

In Yankee Stadium, Carney covers the games with six cameras, and

each camera has several functions.

Camera 1 is high on the third-base side, working from a large basket attached to the mezzanine, and its primary function is covering plays at first base and following lead runners. It is also the camera used for close-ups or left-handed batters, and, in stadiums with deep right field corners, for coverage of the rightfielder. Carney uses camera 1 to ISO runners on the base paths for instant replays (e.g., during double plays), as well, and for pickup shots, such as the manager in the Yankee dugout.

In Yankee Stadium, camera 2 is in the press box, again at mezzanine level, directly above home plate, for coverage of the entire field. During

the game, camera 2 follows the ball. (According to Carney, every sport requires one camera to follow the ball.) Before and after the game, camera 2 swings around to show the sportscasters (usually in front of a chroma-key wall, so the stadium can be inserted behind them).

Camera 3 is the reverse of camera 1; it is also suspended in a basket, high above first base. It is used to cover the infield, to show a spectacular play by the shortstop, second or third baseman, the throw to first base, any plays at third base, and, if the stadium has a blind left field corner, left field as well.

Camera 4 is the centerfield camera, and it is used, like camera 2, to cover the ball. Positioned just

above field level, usually among the bleachers, camera 4 takes the shot of the battery, with pitcher in the foreground and catcher in the background. With its direct view of the catcher, it is also used to cover the catcher playing foul balls (this, like so many plays, can be covered by almost any camera—the question is always "Which camera has the best shot?")

Camera 5 is a dugout camera, located at field level on the outfield side of the visitor's dugout (first base side), gets close-ups of the pitcher, of right-handed batters, and follows runners on the base path. It is also important as a "color camera," picking up details of a play, or the antics of a fan or a team mascot

(once again, any camera can be used as a "color camera," depending upon where the action occurs and which camera has the best shot). Like camera 3, which is mounted above, camera 5 can shoot into the opposing team's dugout.

Camera 6 is the reverse of camera 5, also a dugout camera, and it gets close-ups of the pitcher, close-ups of left-handed batters, and also covers runners on the base path. Some directors prefer to use camera 6 (or an additional camera 7) in a cage directly behind home plate, where it can dolly left and right to cover a wider range of action.

Any of the cameras can be used for color, and any can be used on ISO for an instant replay. Carney has

some definite ideas on the proper use of instant replays. "The most important thing about sports coverage is showing the plays, live, as they happen. That's what the excitement is all about. Some directors get so involved in showing a play from forty-five different angles that they miss the most important shot—the one that shows the action as it happens." Carney works his crew hard to *get the shot,* and then uses the instant replay (which is a camera on ISO to a 1-inch machine, recued and replayed only seconds after the play) to give the viewer more information about the play. Carney is careful not to overuse the instant replay; he elects to show the replay only when it provides the viewer with something new.

Instant replays are part of what Carney calls a "first-class look," a manner of coverage defined by the networks during the 1970s and 1980s. Slick graphics and animated openings are now an important part of coverage, even at the local station level, and so are the details that give a production a "first-class sound"— microphones strategically positioned to pick up bat cracks, crowd noises, the stadium PA system, and, of course, music in and out of each commercial. Network announcers depend upon writing and statistical staffs for their material; with one hundred-plus games to the season, there is no time to prepare a script, so Carney depends upon Rizzuto's indefatigable wit, and his own ability to feed the announcers a line when they need one instead. The result is a distinctive style of coverage that most Yankee fans find colorful, entertaining, and fun. Everybody plays it loose, everybody has a good time.

Baseball coverage is

entertainment, but it is also a business. Carney, though employed by WPIX, serves two masters: the station and the team. WPIX pays the Yankees millions of dollars annually for the right to cover the games, making its money on profits from advertising revenues. The team sees baseball coverage as publicity, a means to promote the team and to sell tickets. Carney must be objective as he covers the action, but the choice of a shot and the amount of time that the shot remains on the screen is a subjective decision. The difference between "objectively covering a error" and "making the shortstop look bad" is always a judgment call. After these many years, Carney has developed a fine relationship with the Yankee organization (which is owned by George Steinbrenner, a man who changes managers and team members with some regularity). The team and the station trust him to make the right decision.

As sports coverage has become more complicated, the combination of producer-director has become somewhat unusual. Carney has grown with the game, with the coverage, with the changes. But it is clear that when Carney finally leaves the game, he will not be replaced by one person. Instead, WPIX will hire two, and maybe even three, people: an executive producer, to handle the management functions and the dealings with Yankee management, a line producer, to take care of the details, and, of course, a director.

stretched to the limit; the idea here is to finish up and go home).

Aftershow get-togethers are common (the first round of drinks is on the producer or director). Winding down together is the best part of the day; it's a time to put everything into perspective, to relax and have a beer. Television production is a team effort; this is the time for the producers, the directors, the performers to be "one of the guys."

Planning for the Edit

Event coverage is most often live, but some multicamera remotes are produced for later viewing. Taped events are rarely shown as-is; they are almost always edited. At the most basic level, editing permits the director to eliminate the errors, to replace bad shots, to fix the sound if any problems occurred. If the event runs long, it can be edited to fit into a fixed time slot.

More often, however, the editing process involves far more than cleanups and tightening. Editing, as described in the next chapter, can dramatically alter the original presentation, create excitement when there was none, and change the whole feeling of the live event.

This is usually accomplished in two ways: by isolating as many of the video feeds as possible and by isolating as many of the audio feeds as possible. With the production stripped down to individual elements, a director can rebuild the entire presentation. This is frequently done at televised rock concerts.

1. ISO

On the video side, isolated or "ISO" camera feeds are common. The director uses most, but not all, of the cameras to cover the action. The ISO camera (or cameras), frequently used in the close-up of a lead singer or other main character, or members of the audience for later use as enhancements or as cover shots in editing, is not used in the basic coverage. Instead, it feeds a separate tape machine (called the ISO machine), whose time codes match the codes on the master tape. By running the master and the ISO reels simultaneously in the edit room, the director can cut shots from the ISO reel into the master. The process works well, provided that the director *never* uses the ISO camera for a master shot.

Any number of camera-recorder combinations can be used for ISOs, with the understanding that fewer cameras will then be available for coverage, and, consequently, more of the work will be done in the editing room. This situation is preferable for many directors, because there is no real pressure in the editing room, none of the angst that accompanies live production, and because there is always the opportunity to run the tapes one more time and "try it again."

The ISO system is also used on live productions either for instant replays, or on larger projects, where several cameras are isolated and one composite feed appears on the director's monitors. An AD usually directs the ISO feed, which is generally coverage of a specific portion of the event, and not the event itself.

2. Isolating the Audio Feeds
Isolated feeds are also the basis for quality audio recording. Musical performances are best covered with not only a video mobile van, but an audio van as well (remote audio vans are most often used to tape concerts for release on record albums). The audio van consists of two principal parts: the multitrack audio mixing console and the multitrack audio tape recorder. The feed from each microphone is assigned to a track, balanced (using a VU meter built into the console), and recorded on one of four, eight, sixteen, twenty-four, or more tracks. The audio engineer takes careful notes, making certain to list the microphone (and performer and instrument) recorded on each track, along with other information that may be helpful later.

Time code, fed to the multitrack recorder by a time code generator in the video truck, is recorded on one of the empty tracks. This code should match the code on the videotapes, so that all of the tapes can be played in perfect synchronization.

The audio tracks are subsequently mixed and equalized (certain frequencies emphasized, others de-emphasized). Most audio engineers make a "rough mix" during the show and record it on one of the empty tracks. A final mix is created by balancing the sounds on each of the tracks. This audio work is usually done independently of the video editing; the tracks are mixed, but are usually edited to match the picture (more about this in the next chapter).

3. Handling the Tape Masters
As all of the tapes return home, someone must be assigned to check them into the library, with proper labels (the name of the production, the date, and whether the tape is a program master or an ISO from a specific camera). If ¾-inch dubs were not made on the truck (some trucks have both 1-inch and ¾-inch machines, but most have only 1-inch recorders), they must be ordered so that the screening process may begin. Screening dubs should be recorded with visual time code (they're called "window dubs," because the time code appears in a window on the screen). If there is only one copy of each camera or program master, backup copies should be made at the same time in case anything happens to the originals.

If the tapes are to be shipped back home, be sure they are carefully packed and that they are insured for the maximum allowable amount (e.g., the cost of remaking the tape from scratch, if the shipper's insurance policy allows). It's usually best to carry the masters by hand, to take them on the plane as hand baggage, to usher them past customs if you're working in a foreign country (and be sure to check the customs requirements well in advance of the flight home). If the tapes have been recorded in the foreign PAL or SECAM system, they must be transferred in a specially equipped facility, converted to the American NTSC standard, because PAL and SECAM tapes will not play on NTSC equipment (and vice versa).

When the tapes return home, each one should be screened and logged as the first step toward planning the edit.

Cleanup

While the director (or producer, or postproduction supervisor) is preparing for the edit, members of the production staff are left to tie up loose ends.

1. Creating the Show File

All of the show materials, including the script, the schedules, the edit notes, the budget, the signed release forms, the insurance forms, all of the important paperwork from the production, should be carefully filed away for future reference. This should be done after every production or series, regardless of whether it is shot on location or in a studio. Release forms (see chapter 10) should be double-checked for signatures and then sent to the proper parties. The producer should make certain that thank-you letters are written to everyone who made the project a success. (Sometimes, a thank-you letter is more effective if accompanied by a small gift.)

2. The "Postmortem"

If the production had some problems, a meeting of all key people should be organized by the producer or director. In order to insure that the meeting is constructive, it is wise to develop an agenda and to bring along a videocassette for screening specific sequences (these meetings can become finger-pointing shouting matches, so it is especially important to stick to the agenda and to maintain an extremely professional demeanor). Most problems can be solved by improved planning (more comprehensive location surveys, more time to set up, more time to rehearse, more gofers at the location site, and, usually, either a larger budget or a rethinking of the way that money is spent). Problems caused by equipment downtime should be reflected in an adjusted bill from the facility or rental house. Some problems, however, can be solved only by replacing the people involved.

Once the event is over, the pressure disappears. The strains brought about by too many hours are history, and even the most explosive situations are forgotten. Nobody bothers to remember the problems; most people tend to remember the fun of working together, the stories about being caught in the rain, and to think of this kind of production as something akin to an adventure.

EIGHT

Postproduction

Postproduction is frequently used as a synonym for videotape editing, but editing is only one of the many steps in completing a production. On any given project, postproduction may involve some or all of the following steps:

1. Screening and logging camera masters
2. Creating an edit plan or paper cut
3. Off-line editing or rough cutting
4. The creation of special-effects sequences
5. On-line editing or assembly
6. The creation of an edited master
7. Laydown, audio editing, and layback
8. Revising the edited master
9. Creation of the final cut
10. Delivery

The importance of each of these steps and the relative amount of time allotted to each depend on the project's requirements, the schedule and delivery date, and, of course, the budget. On a commercial, the work is frequently meticulous, with every picture, every sound, every title, every effect tooled until perfect. On an hour-long industrial, the postproduction work must be professional, but ultimate use, reflected in the budget, may not demand perfection. Talk shows are usually aired as recorded, with editing used to shorten productions that ran long in the studio, or to cut out profanity, etc. Sitcoms usually are recorded two or three times, and one composite is created, by editing, for use on the air. Dramatic programs, which are typically shot with only one camera, are put together completely in the editing room. Every situation is different, and it is the job of the producer to develop the best possible situation for each project. When money is short, projects tend to be presented very much as they were originally recorded (this is the reason for the freshness and spontaneity of programs like "Late Night with David Letterman" and "Saturday Night Live"); when budgets are high, perfection becomes a goal. With perfection, however, comes a tendency to edit too carefully, and the possibility of eliminating the project's fresh and lively approach (this is frequently the case on contemporary variety shows produced in California).

The networks, local television stations, and most corporate installations have their own editing facilities, and, in the larger operations, their own special-effects equipment as well.

Some independent producers, directors, and production companies have invested in their own off-line editing equipment. In the larger cities, off-line and on-line editing facilities can be rented by the hour. Off-line costs about $100 to $150 per hour or so, on-line costs $300 to $500 per hour, depending upon the facility, the equipment required, and the producer's ability to strike a "package deal." Special-effects equipment is also available for rental at these facilities.

Regardless of where the editing is actually done, the process always begins in an office or a screening room, with a ¾-inch machine, a pile of cassettes, and a few dozen blank tape logs (or legal pads, in a pinch).

SCREENING AND LOGGING CAMERA MASTERS

1. A Brief Review of Time Codes

Videotape recorded for later editing should be striped with time code on one of the two audio tracks (on ¾-inch tape) or on the time code track (on 1-inch tape). These codes provide a numerical reference for both audio and video throughout the tape.

Time codes can be laid down in four different ways. If the tape is recording continuously, as it would on an interview, the code should begin at 00:00:00:00, and run until the tape ends. More often, the video recorder is stopped and restarted, but the time code clock runs continuously, so the code may begin at 00:00:00:00, but it may pick up again at 00:03:10:40 (the time that the machine is restarted after a stop), and pick up again after a stop much later, at 00:11:58:24. In the third situation, code may be laid down continuously, but it may start at an odd time (e.g., the

actual time of day, 03:34:28:12). Or it may start at the time of day, stop, and then pick up at the time when the recording begins again.

If the editing involves more than one or two reels, new codes starting at 00:00:00:00, and running continuously to the end of each reel, should be recorded onto each reel before the editing process begins (the procedure, called "striping," records a new "stripe" of continuous time code over the old discontinuous one). Time code can be generated by most 1-inch machines (and by accessory generators, for use with ¾-inch machines), and recorded over old time code tracks. The only problem is this: if the original tape logs were written with the old time codes, they must be rewritten by either creating new tape logs by screening, or by making the corrections arithmetically (if the "offset" between the old codes and the new ones is relatively simple—precisely 4 hours and 20 minutes difference, for example).

2. Window Dubs and Time Code Readers

Time codes can be seen on a special electronic reader connected directly to the videocassette recorder. They can also be "burned" into a window on specially recorded cassettes that can be screened without a reader. Once again, if the edit involves only a few reels, the reader is fine. If there are more than two or three cassettes, the window dubs are preferable (they're also called "viz-coded cassettes," an abbreviaton for "visually coded cassettes"). Window dubs are typically used for paper cutting and off-line editing, because the time code on the screen allows the editor and the producer to identify the source of each edit bite while screening, and to make changes accordingly. When the cassettes themselves are the masters, as in local news reporting or magazine programs, editing is done with a time code reader.

Time codes have become the standard because they provide a dependable numerical index to the contents of the tape, because the computerized editing systems can read these codes and identify specific spots on the tapes by keyboard entry, and because an off-line edit can be made to create a set of time codes that can then be read by a compatible on-line system, saving considerable time during the most expensive part of postproduction.

3. Control Track Editing

Some facilities do not use time code systems because of the expense involved. In these situations, videotapes are recorded with a continuous "control track," an evenly spaced series of pulses that can be read by many editing systems. The control track does not involve any kind of numbering system, so tape logging depends entirely upon scene and take numbers and upon timings taken with a stopwatch. Editing with time codes is frame-accurate (e.g., accurate within 1/30 of a second), whereas control track editing is accurate only within five frames (plus or minus one-sixth of a second). Control track editing requires more work for the producer, and for the editor, so it is less popular. Still, it can be a less expensive alternative to editing with a time code system. The lack of frame accuracy, although frequently used as an argument against control track editing, is not a factor in most productions.

4. Logging Tapes

The tape logs shown in chapter 5 are prepared principally for editing. It is best to prepare these logs in the field and to check them by reviewing the notes while screening the tapes upon return to the office. If no logs were prepared in the field, find a quiet place and watch every tape carefully, noting the acceptable takes and their time codes (or, in the case of control track editing, acceptable takes and their approximate time into the tape). The task can be tiresome, but screening time can and should be used not only to watch, but to plan, noting, for example, specific sounds or pictures that may be useful in telling the story, or as cover shots.

Tape logs should be meticulously prepared with as much information as possible. If a PA prepares the tape logs, as is often the case on larger productions, the director should be able to read and understand them without the PA in attendance. Write clearly, make sure that the descriptions are truly descriptive, and that the takes and timings are absolutely correct. One copy of the tape log should go in the box with the tape, another should be kept in a safe file, and a third should be used as a worksheet for the edit. And be sure to label each log with the name of the production and with the number of the tape.

5. Transcripts

When working with interview footage, some time can be saved by preparing a verbatim transcription of the interview footage, and noting the time codes of key speeches. Once again, the process is time-consuming, and, once again, the time spent transcribing is more useful if it is also used to start planning the edit.

CREATING AN EDIT PLAN OR PAPER CUT

After all of the tapes are logged, the producer, director, field producer, associate producer, or postproduction supervisor can start planning the edit. The goal is to set up the entire edit on paper, listing every take, every sound bite,

Fundamentals of Editing

Although the art and craft of videotape editing can become quite complex, there are only a few basic editing techniques. When combined with one another, and with special effects, the result can be dazzling. A clever team of producer (or director) and editor can make a shaky project look good, and a good project look great. Generally, a good editor is the more important member of the team; he or she knows, almost instinctively as a result of experience, all of the tricks, all of the techniques, and most of the best ways to solve the problems that are bound to arise. With all of the technology, however, there really are a limited number of basic editing techniques, and almost everything is built by using them.

Assembly and Insert Editing

Videotape editing, in its most simple form, is the process of dubbing sound and pictures from one or more camera masters or special-effects masters to an edited master tape. This can be done in sequence, which is usually called "assembly editing," or by placing timed blocks in specified positions on the master, and then inserting one piece in between two others. This latter process, called "insert editing," requires a master that has been "stripped" with either control track or time code from beginning to end, which the machines use as a technical reference.

The Three Basic Edits

Most videotape editing involves the dubbing of both audio and video

from one machine to another. It is possible, however, to dub only the audio (e.g., an "audio only edit") or only the video ("video only edit").

"L Cuts"

The "L cut" is a variation on the three basic edits. An "L cut" usually begins with combined audio and video, and continues as either an "audio only" or a "video only" edit. The term comes from paper cutting, where both audio and video may be represented as a continuous line, and audio-only or video-only is represented as a continuation of that line at half thickness. The resulting pattern resembles the letter "L."

A reverse "L cut" is also possible, where either video or audio leads and the missing track are punched in on cue. With computerized editing, it is possible to mark the exact time of the punch-in, and to have the machine "press the record button" automatically.

Fix-Ups and Pull-Ups

When a tape is edited, the sequence of its recorded parts cannot be altered without rerecording individual elements. It is sometimes possible to make corrections directly on the master by replacing certain audio and video bites, by inserting new sound and pictures, and by cutting out others. If the program is simply too long (or too short) for its time slot, a new master must be created. The program is dubbed from the old master to the new one, and stopped when the first piece must be dropped out or changed. The second piece is then dubbed onto the new master, and, if the show is being cut, is played earlier than before, so it is said to have been "pulled up." Most programs

run long rather than short, so "pull up" is a fairly common term in editing facilities.

On Continuity

The most valuable advice, from a seasoned editor: "Make it look real." Simple advice, but often overlooked. If the performer is looking to the left in a medium shot, the close-up should also show the performer looking left. If the glass of chocolate milk was on the table in one shot, it should not jump to the little boy's hand in the next. If portions of a scene shot in the morning were reshot in the afternoon, do the shadows fall in different directions? A trained editor will see these problems before they're laid down on the master; a good director, preferably one possessing a good visual memory, can be instrumental if the production is a complicated one. Careful note-taking by a trained PA (television's version of the old Hollywood script girl) will eliminate many of these continuity problems as well.

As a rule, don't ever assume that "the audience will never catch it," because they will. And so will clients, who may lose some respect for your technique if it is sloppy.

New Systems Provide Increased Flexibility

A new concept in videotape editing, introduced as the Montage Picture Processor in 1984, eliminates the problem of creating a master that cannot be easily changed or revised. The picture processor concept involves multiple videotape playback machines (in the Montage, Beta machines with high-speed fast-forward and rewind are used), all keyed to the same time code system. The editor determines the order of sequences to be seen, the time in and time out for each piece of tape for each machine. A list is written, but nothing is recorded. Instead, the machines play their pieces in the prescribed sequence, fast-forwarding and rewinding to preset the next event before it is needed. The producer or director simply watches the program monitor, and asks the editor to shorten or lengthen sequences, and adds new sequences as desired. Since there is no need to rerecord when changes are made, the system can save time when used properly. The Montage also uses a computer printer to create a hard-copy picture of the first frame of each sequence, complete with time codes and reel numbers. This "instant storyboard" approach makes paper cutting easy, and encourages further refinements for more perfect finished products.

Several other picture processing (akin to word processing) concepts are in the works.

---

every graphic, every special effect, every element separately, listing the "in" (the time code, word cue, or visual cue that starts the piece), the "out," and any necessary descriptions. In real life, most, but not all, of these decisions and prep work are completed prior to the start of the edit session. Time is money in an editing room, which is billed by the hour (even if an editing facility is in-house, the project's budget is billed for hourly use).

Creating an edit plan is very much like structuring a script. The story must be told by using a combination of existing material and new material, like new voice-overs written specifically to tie existing elements together, or to amplify a thought or idea.

The best way to create an edit plan is to start with a basic rundown of existing pieces. This forms the foundation for the story. In some cases, there will be problems in video transitions (e.g., a close-up following a close-up from a slightly different angle may look "wrong"), and audio problems (caused by off-beat phrasing, discontinuous music or street noises, etc.). In planning the edit, it's best to deal with the audio problems first, because the audio track usually carries the story line. Solve problems one at a time, breaking larger problem areas into smaller components that can be corrected more easily than the large ones. If an audio edit seems to come at a strange place, can a word be eliminated to make the transition sound right? Can a portion of the story be told through pictures, with a new VO in place of problematic live audio? Can the audio from another part of the story be moved to cover a difficult video sequence? Most experienced editors are familiar with the full range of solutions, and will be helpful if invited to solve a specific problem as the edit plan is being made (budgetary restrictions usually limit the editor's involvement, but this varies with each facility and organization).

Once the audio track is in place, the video problems must be solved as well. If the crew shot "B roll" material, most of the visual difficulties usually can be eased with the insertion of background footage, either as transitional or as illustrative material (e.g., showing the jail, showing the site of the fire while the victim is telling his side of the story). If the "B roll" is limited, occasional reverse shots (e.g., the reporter listening) can be inserted to cover a bad transition.

Since moving short audio and video pieces to cover transitions can be time-consuming, most field producers develop the basic edit plan as they're shooting. Even with the best possible planning, and the best crews and the most fastidious technique, there are problems that become evident in the editing room. Solve as many as possible when working on the edit plan; try to allow some leeway in scheduling and budgets; do the best you can. Television can be an imperfect art.

The completed edit plan shows the first five and the last five words of every existing audio bite, a description of every video bite, and both reel numbers and time codes (or takes or timings, depending on the editing system) for every cut. Transitions should be noted as either cuts, dissolves, wipes, or special effects. And any new material should be scripted, timed, and, preferably, recorded prior to the start of the session. The plan itself should be typed, and a copy of the plan should be given to the editor, along with all of the tapes, records, and any graphics to be added through the title camera, at a short meeting that begins the session.

In some situations, the producer or director of the piece supervises the edit session, actually sitting with the editor and reviewing every cut. In others, the editor creates a rough cut on his or her own, reviews it with the producer or director, makes the changes, and completes the piece.

OFF-LINE EDITING OR "ROUGH CUTTING"

There are two kinds of videotape editing: off-line and on-line. Off-line editing is the tape equivalent of rough cutting, and it uses ¾-inch videotape (and, sometimes, ½-inch VHS or Beta tape). On-line editing is the tape equivalent of fine cutting, and it uses 1-inch videotape (and, sometimes, the older 2-inch tape format). Computers are almost always used to control off-line editing systems, and always used to control on-line systems.

A typical off-line editing room includes three or four machines (¾-inch), an equal number of time-base correctors (necessary for maintaining stable video images from ¾-inch VCRs), and a computerized editing system based on time codes (CMX is one of several popular brands). There is a monitor for each playback machine, a monitor for the recorder (to show the edit as it happens), and, usually, a small video switcher for effects and a small

audio console for simple mixing. The editor typically sits at the computer keyboard, in easy reach of the tape machines and the master tapes. The "client" (e.g., the producer, director) is seated behind the editor, preferably behind a desk (so that the edit plan and the tape notes can be spread out); in smaller facilities, a clipboard on the lap replaces the desk.

Most off-line rooms also include a computer printer (to create a hard-copy list of time codes used in the edit), a black-and-white title camera (for insertion of last-minute graphics, as well as the slate), and a character generator. A pair of high-quality monitor speakers is also desirable, as is a pair of low-quality speakers whose sound mimics the marginal quality of most home TV sets.

Off-Line Editing Room
Most off-line rooms are informal places where a few machines and a controller have been laid on tables or cabinets in hopes of later building a more permanent place. This well-equipped off-line editing room includes, from the left, a character generator keyboard, an eight-track audio mixer, and a switcher (the box behind it controls the digital-effects equipment). The monitor is mounted high, within easy sight of the clients (who work behind the editor). The editor's keyboard and screen are located on the right, beside the rack of ¾-inch tape machines. Assorted monitors, tape storage, switching and picture processing equipment fills the back wall.

A good off-line room offers a wide range of capabilities. At the most basic level on the video side, one VCR will rerecord, in sequence, a series of bites from a second playback VCR, with smooth, even cuts between each bite. Add a second playback unit, and a controller, and it is possible to play one tape, then another, without stopping the record machine. Put these signals through a video switcher, and it is possible to dissolve or wipe from one machine to another. Add a character generator, again passing the signal through the switcher, and it is possible to create opening or closing titles and IDs. Add a title camera through the switcher, and a black-and-white graphic can become a two-color graphic by adding colors generated by the switcher. The only real limitation is in the dubbing: the process of ¾-inch editing itself causes the loss of one generation (because the record machine essentially creates a dub of the playback piece), and it is possible to make only one more dub before the picture degrades so badly as to be unacceptable. This becomes a problem if the masters are already dubs of ¾-inch camera masters. Dubs are frequently used in editing interview stories, where all of the material is recorded on one master tape and dissolves or other effects are required (because you cannot dissolve to and from the same source videotape).

Nearly all local and network news programs and many magazine and information shows (e.g., "Good Morning America") use their off-line rooms to create air masters of field reports. For most educational, industrial, and corporate situations, off-line editing is all that is needed to create a first-class production, albeit one without many effects (because effects are usually created in the on-line rooms). Complete programs for broadcast, particularly on network and pay TV, are usually made in the on-line room, and so are tapes that will be used as duplication masters (home video programs, corporate sales tools that will be dis-tributed to offices throughout the country). Commercials, if not on film, are usually mastered in an on-line room, again to maintain quality throughout the duplication process.

When off-line editing is not an end unto itself, its primary function is to create a rough cut that will later be finished in an on-line editing room. Since the cost of renting an off-line room is roughly one-quarter the price of renting an on-line room, producers and directors use the off-line system to make most of their decisions, and then move to the on-line room once these decisions are final. Rough cuts are typically used by the producers to screen the edit-in-progress for clients, like advertisers and networks. On network situation comedies, typically edited in this fashion, each day's off-line cut is sent to the network for review. Many commercials are put together in rough form, sometimes with storyboard drawings in place of finished dramatic or graphic presentations, for testing. And most documentaries are rough cut—and changed many times—in order to make certain that the story is being told in the most effective way. Making changes during off-line is appropriate; making changes later, in on-line, is expensive and frequently unprofessional. The off-line cut is refined as a result of suggestions and changes until all of the cuts, all of the sound bites, all of the pictures, everything, is presented, and paced, as it will be in the finished product.

Once the off-line edit is done, the revised edit plan, essentially a list of time codes and reel numbers, is brought into the on-line room. In most cases, the on-line "conform" is done by typing each of these codes into the on-line computer, one by one. As computerized editing systems become more sophisticated, the off-line and the on-line computers will communicate effectively, and the process of code entry will be automatic.

THE CREATION OF SPECIAL-EFFECTS SEQUENCES

Although most off-line rooms have some special-effects capabilities, the more sophisticated switchers and digital video-effects devices are most often found in on-line editing rooms. In order to use these devices, material must be recorded (or dubbed onto) 1-inch tape.

The use of special-effects equipment can be costly because the per-hour or per-day cost of the effects devices is usually added to the cost of the editing room itself. In some situations, the on-line effects devices are used to build an exciting opening or closing, or a set of bumpers, to be used throughout the run of a series.

Some producers prefer to build all of the special effects before editing the final cut. Others prefer to build effects as they're needed, in sequence. In either case, the producer must allow a substantial amount of time and money if the effects are to be complicated. Even a simple show open lasting from five to thirty seconds, using the character generator, the switcher, a title camera, and a Quantel, rarely takes less than four hours, and may require two full days or more. It is easy to underestimate the cost of postproduction, particularly in special effects, because every situation is different, and because many of the ideas are abstract until the process of building actually begins. As with all editing, there is a balance between creating a piece that "looks right" and spending a "reasonable amount of time and money."

As computer graphics become more sophisticated, devices are being given their own work areas, with their own operators. The paint box, usually operated by a graphic artist with limited technical training, is usually located in a dedicated control room where images can be recorded on disc or on 1-inch tape for later editing. New computer graphics devices also require considerable support equipment, whose demands for physical space have encouraged the creation of video graphics suites or laboratories.

ON-LINE EDITING OR "ASSEMBLY"

1. The On-Line Editing Suite

On-line editing systems are used to create complete videotape masters, ready for duplication or for broadcast. Typically, the on-line room includes a computerized editing console, a switcher, a series of video monitors, the controllers for one or more digital-effects devices, a character generator, a pair of monitor speakers, a small audio mixer, and a title camera. Many on-line editing suites—particularly those available for rental—are set up as small living rooms, very comfortable, frequently with couches and relaxed lighting and some snacks when a session runs late into the night. The 1-inch (or 2-inch) tape machines are located in a nearby machine room; in the larger facilities, ten, twenty, or more tape machines are all located in one room, and assigned to the editing suites as needed.

2. The Process

On-line computerized editing is, for the most part, similar to the off-line editing. Tapes are loaded onto record and playback machines. The editor enters time codes to start and stop specific playback pieces, and gradually builds a master tape. With frame-accurate precision, each edit is scrutinized both by editor and by producer, with split seconds (frames, actually, at thirty per second), shaved off the beginning and end of each sequence to insure a tight, carefully paced finished product. If the off-line work has been done properly, there

On-Line Editing Room
(A) The editor's basic work area, including the keyboard, which allows control, via time code, over all available tape machines, and the screen, which provides a constant update on the edit decision list. (B) Top editor Keny Gutstein at work. (C) Director Bob Small working closely with Keny. The program monitor is mounted up top. The lower-level monitors show feeds from individual tape playback machines and special effects. The audio console is in the foreground. The switcher and a scope are seen just behind Keny, on the left. (D) A wider view of the on-line room shows a pair of monitor speakers (above center), a title camera (left), and the aftermath of a Chinese meal, eaten between 1:00 and 2:00 A.M.

should be no need to cover bad transitions or to change the order of a sequence because it does not play. Still, the addition of special effects does change the way that sequences play, and some changes are usually made during the on-line edit. In some cases, the on-line system may have capabilities beyond those offered by the off-line rig; there are, therefore, situations when the first cut should be made in an on-line room (e.g., laying down a series of small audio bites from a variety of sources can be extremely time-consuming in an off-

line room where only three machines are available; in an on-line situation, where a great many tape machines can be accessed without reel changes, the process is far more efficient). If many machines are available, as is often the case when editing at night or on a weekend, it may be possible to load each reel onto its own tape machine. This saves time-consuming reel changes (which can take as long as ten or fifteen minutes, including rewinding, removing the reel, boxing it, unboxing the next, and so on).

3. Dubbing and Generation Loss

One-inch tape presents no problems in generation loss, so it is possible to create fairly complicated video graphics by using the switcher, the title camera (which may be a color camera in a well-equipped on-line facility), and the digital-effects devices (which may cause the image to degrade slightly if the image is passed through the machine repeatedly). The ability of 1-inch tape to hold its picture quality through multiple dubs is also important when a complicated program is not coming together as planned. Successful components of the program can be dubbed from one reel to another, so the edit need not create an entirely new master each time revisions are made. Instead, the editor simply dubs the portion of the program that is "working" onto a new master tape, lays down new edits, and, when possible, dubs completed sequences onto the new master as well. Ultimately, one must remember that videotape editing is an extremely accurate process of dubbing material from one or more tapes onto a master. If there are changes to be made in that master, the process of dubbing must begin again; nothing can be "moved" on the master tape without rerecording. (See sidebar for information about computerized systems that allow sequences to be moved without rerecording.)

4. Working with Three Audio Tracks

Both the off-line and the on-line rooms allow both video and audio editing, either as one or as individual tracks. On-line rooms usually offer a greater amount of control, which will be increasingly important as stereo television becomes popular. One-inch tape has three audio tracks; any two can be used (sometimes with machine modification) for stereo recording. In the days before stereo television, one track was used for the voices, and another was used for music and effects (M&E). This system permitted easy translation into foreign language versions: the remaining audio track was used for a new (Spanish) voice track, which, when played back in a balanced mix with the M&E track, would create a foreign language sound track with monaural sound. With stereo television, the process becomes more complicated. The original recording, taped on location in stereo and in English, occupies two tracks, and the M&E track would be the third. But stereo playback introduces the need for stereo M&E tracks, balanced to match the location of the voices onstage (if a door opens and its sound is heard mainly through the left speaker, the person entering the room should be heard from roughly the same location). Setting up these tracks is beyond the capability of most on-line rooms; in these cases (which are becoming increasingly common), a multitrack audio room is used to create the stereo soundtrack, which is subsequently dubbed onto the master videotape. This process is explained in some detail in the next section. (The foreign language version would also be created in the multitrack audio room.)

5. Using the On-Line System to Solve Production Problems

Every production has some problems, some plans that did not materialize, some technical mishaps. A good producer assumes that there will be problems, and plans carefully to eliminate as many of them as possible. The screening and the paper-cutting, the off-line edit and its subsequent reviews and revisions, are all checkpoints; with few exceptions, every problem can be solved by a resourceful producer or director, a responsive facility, and some money left in the contingency line of the budget. The on-line room facility is where most of this "fix-up" work is done (the more complicated audio fix-ups are done in a multitrack room). A close-up that does not exist, for example, can be created (with some loss in image quality) by using a DVE to expand a medium shot. A video sequence that runs a few seconds short can be expanded by the slow-motion feature on a 1-inch machine (and the original audio can be dubbed onto the new, slower video, so that it sounds "right"). An awkward image can be reversed by a DVE. The DVE can also select individual still frames to create a slide-show effect—this can be extremely useful if the physical tape has been damaged, or if there were problems with camera or direction.

On one magic show, the slender threads used for levitation illusion, never noticed during the taping, suddenly became visible in the editing room. The editor cleverly increased the overall black level of the picture, and matched the dark background color with a soft-edged oval-shaped wipe to further obscure the strings (while this system would work today, the paint box provides an even better solution—picking up the background color and using the airbrush to retouch the video, frame by frame, treating each image as a still picture).

6. Selecting an On-Line Facility

Since on-line equipment, with the usual complement of special-effects devices, can be expensive, many facilities offer either rudimentary on-line systems or no on-line equipment at all. Most installations have one or more off-line editing systems, because they're not very expensive to buy and they're relatively easy to maintain. On-line systems are expensive, and they must be carefully maintained. In most major markets, there is at least one postproduction facility that rents on-line editing suites, complete with editor, for $300 to $600 per hour (this figure varies with each market and each facility, depending upon both equipment and reputation).

Although most decisions are made as a result of personal recommendations and referrals, the type of work normally done by the editing facility is the first consideration for choosing a place to work. If the facility is known for editing relatively simple commercials, the introduction of an entire show, particularly one with special-effects requirements, might be asking for trouble. On the other hand, a facility that makes most of its money editing big network specials may slight a small industrial (e.g., offer the services of a less-experienced editor, encourage working at off-hours). A large, fully equipped facility is best when the production is complicated, when there are technical problems that require clever solutions, and when the overall excellence of the presentation is a major issue. A smaller facility is usually less expensive by the hour, and everyone in the facility is likely to be concerned with only one or two major projects, but the problem-solving capabilities may be limited. The quality of a postproduction facility is best judged by meeting with the editor who will be doing the job (this is the usual procedure), by touring the work areas (are they clean? well-organized? professionally run?), and by speaking with people who have edited similar projects in the facility.

THE CREATION OF AN EDITED MASTER

The amount of time required for on-line editing depends entirely upon the complexity of the project. Many network sitcoms do their off-line editing in about two days, and their on-line in the course of one long day. A commercial with many special effects could take as long as a week to complete; a music video takes about as long. Building an hour-long variety show from existing sequences usually takes a week or two. If the pieces have to be created in the editing room (because they were shot single-camera, for example), editing an hour-long show usually takes the better part of a month, or even longer in special situations.

The time is spent making each cut, each transition, each special effect as perfect as possible within budget limitations. After a period of time, the project starts looking like a complete production, and can be screened for casual observers (this can be helpful in determining the proper pacing, the overall clarity of the presentation, and, in a comedy program, where the laughs really are). After a reasonable amount of tinkering and fixing up, the edited video master is complete. If there are no further audio requirements, the program can be prepared for delivery. If a sound track must be prepared, there is still a few days' work to be done. In either case, the edited master should be dubbed to create a "protection master," which goes directly to the library for safe storage. In order to minimize the number of passes on the master (a run through the machine is usually called a "pass," and each pass strips some of the image-bearing magnetic particles off the tape, and risks some accidental physical damage to the tape, so passes on the master are to be kept to a minimum), any ¾-inch screening dubs should be made as the protection master is being made. Or further dubs should be made from the protection master, and not from the edited master itself.

LAYDOWN, AUDIO EDITING, AND LAYBACK

When a production's audio requirements are beyond the capability of the videotape editing room, the sound track must be prepared in a specially equipped audio mixing room. On a 1-inch VTR, three tracks are available with limited opportunity for manipulation between them. On a multitrack audio recorder, at least eight tracks are available, with twenty-four tracks common, and thirty-six tracks used on the more sophisticated machines.

Why are so many tracks necessary for a production whose final sound track will most likely be heard through the low-fidelity monaural sound system common in today's television sets? Because television sets are being designed for better sound (many with stereo capability), and because a good, clean track will always sound better than a sloppy one, even through small speakers. With multitrack recording, each sound source is recorded on its own track, so it can be modified, edited, and perfectly balanced with each of the other tracks to create a sound track that sounds natural. Just as only the most skillful of audio wizards can capture the vitality of a symphony orchestra with the absolutely precise placement of a single microphone, only the most skillful television audio engineer can capture picture-perfect live sound. The rest of us must create the best possible track by combining individual pieces.

1. The Audio Room

An audio mixing room for television combines many of the aspects of standard recording studio design with the necessities of a video editing system. On the audio side, there is a large multitrack mixing console, complete with faders, equalizing circuits, and accurate metering of each track, as well as two pairs of speakers (one pair of extremely high fidelity, and another, much smaller, pair whose fidelity resembles that of a standard television speaker), one or more multitrack recorders, and a complement of record/playback equipment (e.g., audio cart machines, four-track recorders, audiocassette recorders). The room is carefully soundproofed, designed for the best possible reproduction of sound.

On the video side, there is either a 1-inch

Audio Room

(A) Most editing rooms are equipped with a relatively simple mixer. This six-channel unit allows control over volume (by using the faders, bottom left), equalization (top left), and balance between stereo channels and the monitor speakers (below the left meter). (B) For more sophisticated control, this thirty-six track room allows considerably more flexibility in equalization, in playback and recording individual tracks, and in time code control of the multi-track machines in coordination with picture. Note the small square speakers mounted on either side of the console; they are used to hear the sound track as it will be heard on home receivers.

VTR for playback (and subsequent rerecording) of the sound track or patch-panel access to a machine located elsewhere in the facility. A ¾-inch machine, with time code capability, is tied to the time code system in the multitrack audio machine so that sound and picture can be played simultaneously (this is the key to the whole audio/video editing process—audio and video must lock up in perfect synchronization). A color monitor, frequently a large one (either a projection set or an oversized monitor—the large screen is a holdover from audio editing for motion pictures), is usually located in between the high-fidelity monitor speakers.

With both audio and video locked to the same time codes, it is possible to review sequences by hearing one or more of the tracks with the multitrack machine, and by seeing the pictures from the ¾-inch video playback at the same time. Changes can be made on the multitrack, and, as long as the time codes are not altered (and there are no technical mishaps), sound and picture will be forever married.

2. Laying Down the Tracks

The audio editing session begins with a process known as "laydown," where the audio tracks and the time code track are dubbed from the edited video master onto empty tracks on the new audio master. For purposes of illustration, we'll assume that there is only one audio track (e.g., the program has been recorded in mono), which will be recorded onto track no. 2, and that the time code will be recorded onto track no. 23. Tracks no. 1 and no. 24 will be left empty, because these are physically located at the outside edges of the tape and are therefore most susceptible to physical damage through mishandling. Track no. 22 will also be left blank, just in case the beeps and clicks of the time code track bleed over into an adjacent track. Track no. 3 will be used for live sound effects, recorded on location for later placement. Track no. 4 will be used for "room tone" (every room has its own distinctive sound, and a mix without it will sound "wrong"—every location recording should include at least a minute of room tone.

with no sound or activity to clutter it, for later use in postproduction). Track no. 5 will be used for music, again recorded in the proper places to coincide with the action. With these basic tracks in place, the audio engineer can create a "rough mix," and record it on any other track for future reference. Track no. 6 will be used for recorded sound effects, mostly from a sound-effects library available on record albums. Most engineers designate a specific section of the tape for mixes; we'll start our mixes at no. 15, but the matter is quite arbitrary.

The voice-over announcer, hearing the rough mix in a headphone, works in a booth adjacent to the mixing room, watching a playback on a small video monitor. The new VO tracks are laid down on track no. 7.

3. Editing

Just about every sound, every voice-over, every effect is now in place. Before the final mixing begins, however, some audio editing frequently will be required to prepare for the mix. Certain sequences will be short, so additional sound effects, or repeated sections of music must be added in their proper places. Errors in the VO track, pops and clicks during the interviews, all of these must be "cleaned" one by one by adding sounds, by erasing small portions of the tracks to make mixing as simple as possible. Complicated productions usually have long lists of audio edits to be made; even the most carefully planned projects are bound to have some audio problems. The work should be done carefully, with a checklist. There's nothing worse than having to stop a mix because a piece of material is not where it should be, or because it runs too short or too long—everything must stop until the problem is corrected by audio editing.

4. Equalizing

The quality of the sound must be perfect as well. The announcer may sound a little bassy; some of the sound effects, shrill. Most audio consoles can correct these and other related problems by passing each individual channel (up to 24, 36, whatever the capacity of the room) through equalizing circuits that add or subtract certain frequencies, to emphasize or deemphasize highs, lows, mid-ranges, or any other tones in between. To correct the announcer problem, the audio engineer can replay the VO track (no. 7) with some of the bass frequencies deemphasized, and a slight emphasis on the mid-range and higher tones. When the producer agrees that the announcer sounds better, track no. 7 is replayed through the equalizers, and rerecorded as track no. 8. Most of the sound effects are fine, but the whistles and the fire trucks are irritating. When these sounds are "in the clear," without other sounds that will be affected by dropping out some of the highs, the change can be made by rerecording portions of track no. 3 onto a new track (no. 9). When they are affected, the producer and engineer decide that the track will be mixed with slightly less gain (volume) than the usual. Track no. 9 now becomes the sound-effects track. And we are ready for the mix.

5. Mixing

Here's the track rundown (which is usually kept on a chart by the audio engineer, and then packed with the master audiotape for later reference). On our list, the useful tracks are capitalized; the old tracks are not.

1. (blank)
2. VIDEO MASTER
3. Live sound effects
4. ROOM TONE

5. MUSIC
6. RECORDED SFX
7. Voice-overs
8. VO/EQ
9. LIVE EFX/EQ
10. (blank)
11. (blank)
12. (blank)
13. (blank)
14. (blank)
15. Rough mix No. 1
16. (blank)
17. (blank)
18. (blank)
19. (blank)
20. (blank)
21. (blank)
22. (blank)
23. TIME CODE
24. (blank)

The new mix will be recorded on track no. 16. With the rough mix (sans VO) already completed as a rehearsal, the engineer will treat this as a final mix. Each of the tracks is balanced and played as a preview until the first problem area creates a natural stopping point. On some productions, an entire half-hour can be mixed with just a few stops, but some thirty-second commercials will require a few dozen for perfect balance. After each stop, the last section is reviewed, and if all is well, the mix continues. If there's anything to be fixed, it is fixed before moving on.

6. Layback

Once the mix is complete, it is best to review the entire production from beginning to end, without distractions, without interruptions. Once approved, the mixed track is dubbed back onto the 1-inch video master, again locked into place by time codes, to complete the project. Since the dubbing (or "layback")

occurs in real time, the process provides an additional opportunity to screen the finished product.

7. Sweetening

Most people know that sitcoms and variety shows are "sweetened"—with laughs added to tell the home viewer what's funny. Sweetening rounds out the audio presentation, makes the show sound complete. When it is used properly, sweetening can make a good show sound even better. Unfortunately, sweetening is used most often to add some feeling of humor to an otherwise lackluster product.

Sweetening is usually done in the editing room, but may be done in the audio room as well. It is essentially a multitrack approach, where the sound track on the master videotape is rerecorded with laughs and other audience reactions. Although the process appears to be relatively simple, there are only a handful of specialists who know how to sweeten a show properly. As you might expect, most of them operate small facilities in the Los Angeles area (but they will travel with a portable sweetening facility, consisting of a small mixer and a library of prerecorded tapes, for a price).

REVISING THE EDITED MASTER

Once the master has its sound track, and the protection master has been made, several additional dubs are ordered for approvals. Few projects are completed only once; most undergo some changes before they are accepted. Revising a master that was created entirely in a video editing room can be tricky, depending on where the cuts must be made (it is difficult to cut into music, for example, unless the cut is made at a natural breakpoint). Revising a tape made in both the video and audio editing rooms certainly provides more technical flexibility (working with music, once again, is easier in the audio room, where the entire music track can be slipped a few seconds in either direction by rerecording it—but the entire program would have to be mixed again to accommodate the change). Changes are best made first in the video room, where the edited master is once again brought to a finished state; the audio room is then used for fix-ups. If changes affect the overall time of a broadcast program, accommodations must be made to again bring the project to its appropriate running length.

It's best to save all of the old masters, carefully labeled, and to save all of the work tapes as well. The new tapes should include rundowns that explain the differences between the old and the new in a way that will be comprehensible to another producer or director two years later.

CREATION OF THE FINAL CUT

Just as there can be only one boss, only one person can ultimately deem a program "complete." This "right of final cut," as the lawyers would refer to it, usually resides with a senior person in the company responsible for the financing. Network executives have final cut on programs produced for their networks; clients have final cut on commercials produced by individuals and by advertising agencies; producers and directors, unless they are particularly powerful, rarely have final cut on their own projects.

Since there is a direct relationship between the right of final cut and the ability to determine what will—and will not—appear in the finished production, the right to determine final cut is not granted lightly. It has everything to do with creative control, and may

even have free speech implications on some documentaries.

DELIVERY

With the final cut approved, the production must be prepared for delivery. When a production is made by a contractor (e.g., an independent producer or production company), a list of "delivery requirements" is usually included in the contract. The list is likely to include the following: at least one master videotape suitable for broadcast, made in accordance with technical specifications attached to the agreement; the actual running time of the program; a complete script; a complete list of credits; copies of all production agreements (e.g., talent agreements); still photographs for use in advertising and marketing; a complete budget report (this is usually submitted within sixty or ninety days of the official delivery date). Meeting the delivery date and fulfilling the delivery requirements are important; final payments are usually keyed to the delivery and acceptance of all materials.

When a production is made in-house, the tape usually is submitted along with certain information needed for broadcast or for duplication. The tape should be packed with a rundown, with an accurate running time clearly marked on both the tape (either reel or cassette) and on the tape box. Agreements, credit lists, budget information, all relevant paperwork should be copied and placed in a file for later reference. If agreements will have future implications (e.g., royalties paid if the production is sold to ancillary markets like syndication or home video), they should be filed with the appropriate legal, business, or accounting departments.

Every producer should make a habit of creating "show files" at the conclusion of every production. The file should include scripts, notes, contacts lists, timing sheets, rundowns, agreements, notes from the development process, old proposals, anything that might prove useful in the future. One copy of the file should remain in the office. Another should be filed in a safe place for personal use. DON'T THROW ANYTHING AWAY! Five, even ten, years later, a complete show file can save weeks or months of research and development work on a subsequent project (these same files can protect you if any legal problems arise as to clearances, ownership of materials used in the production, libel, slander, and so forth).

Be sure to keep a dub of each project for your own use as well. Many facilities do not permit producers or directors to make dubs.

Even if you must pay for a dub from your own pocket, make sure that you keep copies of every bit of work you do; having a copy of an old project could be important in landing a job. And don't assume that copies safely stored in a videotape library are forever secure—librarians make mistakes, tapes are erased for stock and to make room for the new, and, of course, company policies change every few years. The only way to be sure that you'll have a copy of your work is to own that copy and to place it, securely, in a controlled environment, in the safety of your own home. The very last step in producing a program, a commercial, a music video, or any other project is to make a copy of the finished product for your own use. Copies on ¾-inch tape are best; VHS and Beta dubs, except in special cases, are amateurish.

NINE

Assembling the Team

New producers tend to think that they must be able to do it all; experienced producers understand that their role is one of guidance and supervision. Television production is always a group effort, with the producer the person in charge. He or she makes the entire cast and crew feel that it's their show. The producer must maintain an overview, and must delegate, and also provide credit where it is due. Success depends, of course, on the producer's ability to assemble the best possible team.

THE PRODUCTION STAFF

After working on a number of projects, most producers start to develop a cadre of favorite associate producers, writers, directors, production assistants, designers, music directors, and other personnel. This list of talented and dependable people is one of the producer's most important tools; the ability to move quickly and efficiently during the stages of a production is frequently crucial to a project's success. Since most television productions are made on a deadline, most producers wisely

depend upon established working relationships.

On many productions, problems with availability or unusual requirements may create new employment opportunities. A producer who has accepted an assignment to produce her first medical program, for example, may choose to hire an experienced medical writer with no television experience to fill an important slot. A producer who has just upped a long-time PA to associate producer is likely to move a secretary to the PA slot, or a newcomer who comes highly recommended.

Because training new production staffers takes time, most producers try to hire people with some experience, even for the most basic jobs. Newcomers are most likely to be hired if they possess something beyond college experience, like a background or talent for writing, or a good recommendation, or unusual expertise (e.g., on a children's program, experience with kids is a plus).

Most productions require relatively small staffs. Some producers work with an extremely structured system, where each person has one job and is expected to limit activities to it. I personally prefer a system where each staff member has specific responsibilities but can cover for the others in case of vacation or illness. How a producer works a staff is a matter of personal style.

WRITERS

Every television production has some kind of script. Some projects are obviously scripted, like situation comedies and news programs. Others appear to be ad-libbed, principally because the script has been written in an informal fashion; the VJs on MTV are reading from a script, although they do ad-lib in some situations. Most interviews are scripted, at least in part. Even network coverage of baseball and football games is scripted (and many of the apparent ad-libs, particularly those involving statistics or history, are researched and carefully worded by a writer just outside the booth).

1. Specialization

Television writers tend to specialize. Some are particularly good at story work, and can manage effectively in either drama or comedy. Those who are funny specialize in sitcoms and variety shows, and other types of entertainment programs. Writers who do commercials must be very good at saying a great deal in a very short span of time (this is a *very* specialized talent, whose practitioners may be extremely well paid). Journalistic writers may be equally at home in either news or documentaries; many have newspaper or magazine training. There are specialists in medical, legal, and financial writing. Writing for a game show requires a special set of skills (and, even here, there are specialists in hard-core Q&A, and specialists in humorous material). Most working writers are skilled in several disciplines.

2. Selecting a Writer

Selecting a writer, or a group of writers, for a television project is largely a matter of reading a writer's sample scripts and spending some informal time with the writer to understand how his or her mind works. In many cases, the writer can only be judged by sample material created according to the producer's instructions, on assignment. The writer is usually paid a small fee for this test, but "writing on spec" (speculation) is fairly common, especially in situations outside the jurisdiction of the WGA.

Most writers are hired because of their credits, recommendations, and their ability to make a positive impression on the producer on the first or second meeting. If the project is a one-shot, the producer will work closely with the writer or writers to create a polished draft, sometimes with the help of second or third writers for specific sequences. In network series and movies made for television, the writer and the producer are often the same person. Because the selection of writers is so difficult, many producers of nonnetwork shows have become writers "because it's easier to do the job myself."

The selection of a writer should not be taken lightly. The writer not only provides the words, but he or she also forms the entire structure, frequently provides the performer(s) with a special style, and contributes greatly to the pace and the overall feel of a production. Considerable time and patience should be taken to insure a perfect (or, at the very least, absolutely acceptable) script. To borrow from the theater, "If it ain't on the page, it ain't on the stage."

3. The Writers Guild of America

Writers deal in ideas. Getting proper credit for those ideas, getting properly compensated for those ideas, maintaining control over the ideas and the words with which the original ideas are expressed is not always easy. The Writers Guild of America, formed to protect screenwriters, has established working rules and rates with the commercial networks and with many independent production companies. With few exceptions, programs produced for ABC, CBS, and NBC are written under WGA agreements, as are all motion pictures produced by the major studios.

The concept of the union shop applies here. Members of the Writers Guild may write only for companies who are signatories to the union agreements. Writers cannot work for companies who are not signatories, and signatories may not hire nonunion writers (unless, of course, the writer is a newcomer who will be signed as the first project reaches fruition).

The WGA protects its members in several different ways. First, it establishes the terms of a contract between writer and producer: the rates and payment terms for treatments, scripts, rewrites, and story lines, and the rules pertaining to co-writing, future uses of the program, royalties, and pension and welfare contributions (the WGA maintains a fund for its members) and a range of other details. Second, it demands that the producer clearly define the role of the writer, and guarantees that the writer will receive proper credit for the work done. Third, it provides a registration service for scripts and treatments, so that a writer who is unsure of a producer's morality can submit a script to the Guild for later use as evidence, if need be. Fourth, it keeps a watch over the entire industry (or at least the part of the industry under its jurisdiction), advising members of problematic producers, offering directories, seminars, and other courtesies common among professional associations.

4. Nonunion Writers

Most of the writing done for television, however, is not covered by WGA agreements. The Guild limits its jurisdiction to major motion pictures, the commercial networks, and some local stations in large television markets. Since the majority of television production is done at the local stations, at corporate facilities, and in educational institutions, most television writers are not protected by the WGA. In these situations, rates are completely negotiable. There are no rules beyond those negotiated by the writer (and most writers at this level are incapable of hiring agents, so they do their own salary or fee negotiations and strike the best possible deal under the circumstances). On a free-lance project, the writer may or may not get paid, the script may be handed to any number of writers for rewrites, and the writer may or may not get credit for his or her work.

Life can be difficult for a television writer in the open market. The successful ones hire lawyers or business managers or even accountants to handle negotiations, but most simply manage to earn a living by taking jobs as they come along, frequently at small fees. Although this situation is not likely to change dramatically, many producers have learned that better-than-average treatment frequently yields better-than-average work, closer attention to deadlines, greater willingness to do rewrites, and an easier path to a finished product. Dealing with a writer, with any creative person, may require some additional patience, and, in some cases, a close working relationship. It is *always* worth going the extra distance to make the writer comfortable, to keep the writer happy. Special attention is very much evident in the finished product.

DIRECTORS

We've already discussed the hiring of directors, and the criteria that many producers use to make their selections. (For review, see chapter 2.)

The Directors Guild of America, or DGA,

claims a jurisdiction similar to the WGA's: the motion picture business, network television, and some local television markets. Like the WGA, the DGA sets rates and rules by negotiating contracts with the networks and with the larger production companies. The DGA agreement takes one further step: when a DGA director steps into a studio, the AD and, in most cases, the stage manager must also be members of the DGA (if a control room PA does some of the work normally associated with the AD, like timings and coordination of videotape recording or playback, he or she must also be a DGA member). On film sets, the production manager must also be a DGA member. Since the director is working with a large number of people, any of whom could cause a problem with the union, most directors are careful to follow the rules. (Some WGA writers earn additional money by using a pen name for nonunion projects; but a director is out in the open for all to see.)

Directors in local television markets, and directors who traditionally work on a contract basis for corporations, the record companies (for music videos), and other markets generally work for fees and under rules that have been established by traditions and by marketplace conditions. Unlike the writer, who usually works alone or with the producer behind closed doors, the director works with a large number of people where contributions are usually apparent. Directors, therefore, have little trouble claiming credit, and, except in isolated instances, they rarely suffer from the vagaries of agreements so common with writers. The director does the job, the project is delivered, and the director gets paid for his or her time.

Some directors have representation (this is the case with higher-priced directors), most do not. Most producers and production organizations readily understand the value of a good director, and are generally willing to pay a premium for talent, ability, experience, and reputation.

PERFORMERS

There are two ways for a performer to become involved with a television production: by sending demo tapes or by auditioning. With the importance of program packages in network, pay TV, and home video, the performer, typically a celebrity, may be integral to the overall production concept.

1. Photos, Demo Tapes, and the Early Selection Process

When a television job becomes available, the producer usually "puts the word out on the street," contacting other producers who may know performers who can do the job, as well as agents and managers in the community. Hiring performers is usually a buyer's market —there are many more performers than there are jobs (although finding "the right face" or "the right personality" may be difficult in some of the more demanding situations, like commercials). Rather than seeing everyone who wants the job, the producer may elect to interview a few of the more promising candidates; most producers prefer to see some of the performer's work prior to the interview. This is most often accomplished by screening demo tapes, usually ¾-inch tapes that run five to ten minutes and feature the performer in three or four different situations that demonstrate the range of his or her talent. An eight-by-ten glossy photograph, with a list of credits, union affiliations, and contact information on the back (stapled to the back is fine; printed is a little better), helps to fix the image of each individual performer in the mind of the producer.

To determine a performer's flexibility, the producer can devise a variety of informal tests. These often show the performer's real personality (as opposed to a stage persona, which some performers can turn on and off like an electric light). He may set up intentional mistakes or improprieties that force the performer to react naturally, for these will reveal a great deal about taste, inventiveness, sense of humor, and "street smarts." During auditions he looks for varying degrees of comfort on camera, the ability to read a prompter with an informal mastery, the telltale ad libs (when performers are off-script, they often reveal quite a bit about personality). In short, he tries to develop an image as a home viewer would.

A producer makes use of all available help, oftentimes asking people who pass by the screening room whether they "like" the performer, what impressions the performer delivers. And during the performance, by glancing over at those who are watching—trying to catch a smile, an appreciative nod, or a wince—a good producer can use subconscious comments to great advantage. Ultimately, instinct will guide the decision. Learning to trust those instincts is an important part of a producer's training.

In local television, there are four basic types of performers: talk show hosts, news anchors, reporters, and sportscasters. In these situations, demo tapes provide a relatively good, and usually quite specific, idea of the performer's talents and abilities. The selection process is usually straightforward; performers (like producers) tend to work their way up from the smaller markets (Tulsa, Syracuse) to the larger ones (Houston, Boston).

In corporate video, much of the work tends to be as spokesperson or interviewer. Once again, previous experience provides an important guide to the performer's ability to deliver a script with credibility and appropriate flair. Most corporate work involves reading a script, while some involves the ability to interview executives, or to moderate a discussion. In these situations, the producer tries to select someone who fits the company image, someone who would seem quite comfortable occupying an office of a vice-president. And he tries to match the age of the performer with image, going for younger performers when the company is innovative, more mature performers when the company is traditional and conservative (like a bank or an insurance company).

In educational video, performance ability is at least as important as credentials. Some teachers are superb performers, every bit as interesting on camera as they are in the classroom. Others, though quite excellent in the classroom, develop a severe case of stage fright when asked to appear on television. Most fall somewhere in between, so the producer must work hard to develop the proper balance. This is sometimes achieved by the video equivalent of "team teaching," where the professional teacher handles some of the material as the resident expert, and a professional television performer, working from a script, does the rest.

On dramatic and comedy programs, demo tapes provide only a limited feeling for the performer's talents. But they do provide the producer with an indication of "type," and this is useful for limiting the number of people who must be auditioned.

2. On Working with Amateur Performers

Ultimately, the quality of a television production depends almost entirely upon the talent, the performers. Even with the best possible special effects, the most expensive sets, the best technical crew, the most talented writers—if the performers are not effective, people will not want to watch the program (or commercial, etc.).

Fortunately, there is a large supply of talented performers who make producers and directors look good (actually, this flows both ways). But what happens when a producer or director is faced with an amateur: someone who does not normally appear on television, someone who may even be uncomfortable with the whole idea.

If time permits, almost anybody's performance can be improved by spending hours in the studio reviewing tapes, with the director offering specific instructions regarding movement, gestures, pacing, reading or conversational style, and overall demeanor. Most amateurs are intimidated by cameras—they're not sure whether to look directly into the camera or to "just talk" and allow the camera to cover the action (the latter is generally best).

Finally, a novice can be made to look better by some intelligent visual editing—cutting away to the audience, to other guests, to graphics, to a wide shot, to absolve the newcomer from some of the responsibility for "carrying" the show.

3. Auditions, Call-Backs, and Decisions

The only way to truly judge a performer's ability to handle a specific job is to watch that performer work with the script, preferably on camera, preferably interacting with other performers. While desirable, on-camera auditions are usually reserved for the final decisions. More often, a group of performers, some selected from demo tapes, some suggested by other producers, some who managed to hear about the job and just "showed up," sit outside an office or rehearsal hall, while the producer, director, and, frequently, a program executive briefly interview each candidate, and then request a reading from a portion of the script. Each interview should begin with a question about availability ("We're planning to shoot this commercial in Antigua on October 16, we're going to leave on October 14 and return on October 17. . . . Are you available on those dates?"). Performers who are not available need not apply.

There may be as few as three or four performers waiting in the hall, but on larger entertainment programs, there may be a hundred or more (particularly when the casting person has issued an "open call," letting agents know basic types, and then inviting anyone who feels that they fit the bill to show up for an interview or audition). Being the producer, walking past all of these hopefuls, can bring on a rush of ego in new producers. The more experienced ones know that most of the day will be spent with performers who are totally wrong for the part, who don't have the look, don't have the experience, have the look but don't have the voice, who are trying too hard or not trying hard enough. Once in a while, a miracle occurs, but most of the work, particularly when there's an open call, is an unfortunate combination of tension and drudgery.

The few performers who fit the bill are invited to come back and audition, either on camera or onstage (depending upon the production's requirements and the available facilities). These auditions can be quite extensive, especially when a number of performers are asked to interact with one another to give the producer a sense of the chemistry. After some time, favorites emerge and choices become limited to a handful of performers.

There is no need to make the final decisions at the auditions or at the call-backs (when the best performers are asked to return and read again). It is best to thank everyone, indicate some general preferences if appropriate, and then to regroup with director, program executive, casting director, whoever has been involved from the start. The best performers are offered jobs; if they turn the offers down, the second bests are contacted *as if they were first choices* (this is the reason why performers should not be told that they do or don't have the job at the auditions; it limits the producer's flexibility in negotiation).

When camera facilities are available for tests, scripts should be prepared to show, in a relatively short time, the performer's ability to handle every possible situation. If the production involves reading news copy from a prompter, interviewing guests, and ad-libbing a closing, then each of these should be included in the audition script. Decision-makers should be relegated to the control room, or to a screening room, and should refrain from making any comments in the studio, any comments in the green room, any comments anywhere near the performers. Decisions based on camera tests are rarely made immediately; there are usually a number of executives who will want to see the tape first. And, remember, the performance on screen is what matters, regardless of "how it felt" in the studio.

4. Making the Deal

Most talent contracts are negotiable, with no established base figure and no established limits. In commercials for national or large-scale regional distribution, network and pay TV (and in the movies), the performers' unions (SAG and AFTRA, both fully described in the next chapter) set minimum rates and establish rules. This is true in some of the larger local television markets as well.

In most cases, the marketplace establishes basic rates, and the performer develops his or her own price per day of shooting. If the performer is experienced and well known, as is the case for some of the busier performers in national commercials, the rate can be as high as several thousand dollars per day. If the performer is not particularly well known, and there are no union rates in the market (e.g., corporate video), the fee is usually set by the producer, and may or may not be negotiable.

Some performers negotiate on their own behalf. The more experienced performers usually pay 10 or 15 percent of their fees to an agent, lawyer, or manager who negotiates for them. (Producers in smaller markets may be put off by this behavior, but it is appropriate in just about every situation. Performers should concentrate on performance; in rare cases is a good performer an equally good businessperson.)

5. American Federation of Television and Radio Artists (AFTRA)

A performers' union with jurisdiction over network and big-market local television, AFTRA (and its sister union, the Screen Actors Guild, or SAG) allows its members to work in cable and pay TV and in home video with special permission. AFTRA sets minimum rates and residuals for replays, collects pension and welfare on behalf of its members, and sets basic rules regarding expenses, work away from home, wardrobe allowances, and so forth.

Also common in television are three staging and technical unions: the International Alliance of Theatrical and Stage Employees (IATSE), the International Brotherhood of Electrical Workers (IBEW), and the National Association of Broadcast Employees and Technicians (NABET).

SET DESIGNERS

The "look" of a television production is an extremely important element, one that is frequently minimized on smaller-scale projects. The set (and the music, which is discussed later in this chapter) gives a production its overall quality and much of its unique flavor. A good set provides far more than a place to sit, or a backdrop; it is an essential visual statement, an element of the production every bit as important as the script, the performers, and the direction of the project.

1. Form Follows Function

The design of a set begins with the producer, or director, establishing the functions that the scenery must fulfill, its basic working elements. On a talk show, for example, there must be a place for the host to sit (frequently behind some sort of desk), and a place for the guests. On a dramatic production, or a commercial, the script will provide basic clues about functional elements (e.g., the fireplace, the sofa, the picture window). On a game show, whose sets are among the most functional in television, there must be a gameboard, a place for the contestants that offers a full view of that gameboard, easily readable scoring units, and a variety of other special elements. On a dance program, the only functional element may be a good floor.

On most productions, the design process begins with the producer's explanation of basic functional elements. The art director or scenic designer (the terms are used interchangeably, but may have distinct meanings in larger productions) then offers some basic ideas about ambience: the look of the set, the feeling, the colors, the textures, the relative sizes of the various scenic elements. Some designers can do some rough pencil sketches on the spot, but most prefer to work at a drawing table for a few days to develop concrete ideas. Budget, studio dimensions, and storage requirements should be discussed before the designer starts to work, because the set must be designed within these parameters.

2. Reviewing the Initial Designs

The designer's first submission provides the producer with the earliest visualization of a project's overall look, and some time should be taken to discuss every element in the design in detail. The design must function as a single unit, providing the director with a visual means of tying the production together. Individual elements must function as required by the format, with the various staging areas conveniently located to permit easy movement of performers, props, and cameras. The design motifs must fulfill a number of different requirements. First, they must be functional. Second, they must look good on camera (some materials lose essential details or textures on camera; other surfaces don't look like much in person, but look very good on camera). Third, the elements and motifs must be complementary, so that all of the set elements appear to be "synergistic," all working together for the larger whole. Patterns and colors are one way to tie the whole set together. Shapes and continuing lines are also effective.

Some producers and designers prefer to play it safe, sticking with cool colors, light blues, neutral greens, and grays, working with fairly realistic pieces, designing sets as they might design their living rooms. Others work in a far bolder way, using bright oranges and greens and yellows to enhance a lively production. The set used on "Family Feud," with its homespun shapes and textures, and its bright orange background, combines the traditional with the vibrant, one of many indications that a stage setting should be more than just a reflection of real life. "The Tonight Show Starring Johnny Carson" combines a living room feeling (the desk and interview area) with an aura of big-time show business (the colored drape that provides the background for his monologue; the band area; the luxuriously large picture window). The various "Saturday Night Live" and MTV sets are more impressionistic than realistic, designed more for image than for function. "The CBS Evening News" uses an appropriately conservative setting (essentially a background, a map of the world, in relief, painted in neutral tones), while "ABC World News Tonight," a somewhat flashier show, features the busy newsroom as a key element. The scenic elements on "60 Minutes" are magazine pages; the set is used, very effectively, to say, "We are the television version of a weekly newsmagazine." Cable television's The Weather Channel depends on a large, full-color weather map (inserted from a computer graphics source by using chroma-key) firmly establishes both the function and the feeling of the presentation. Some of these sets are remarkably simple; all are extremely effective.

3. Working with Limited Funds

Every production has limited funds available for its stage setting. On smaller projects, there may be just enough money to buy paint and a few chairs; on larger projects, there may be enough money to build the basic setting, but some of the enhancements are likely to be modified or eliminated for budgetary reasons.

The key to the producer-designer relationship is focus, because focus allows the designer to establish priorities. It is important to concentrate on what the camera—and ultimately the viewer—will actually see. An experienced producer avoids the temptation to create a full-scale stage set with all the trimmings, and concentrates only on elements that will be seen, and seen often. Available money should be used for the set as a whole (or at least for those elements that will be seen most often), not concentrated on one or two particularly desirable set elements. It's easy to fall in love with an especially unusual set piece (like a wrought-iron spiral staircase that goes nowhere, but adds a touch of class), but one must always consider the functional elements first. High-maintenance materials (e.g., painted floors that must be repainted often; chrome or mirrors that require constant polishing) should be used sparingly, with money budgeted for upkeep.

Larger elements usually mean more than smaller ones. And color is frequently more effective than texture (in other words, paint instead of wallpapering to save money). A set that looks sparse on camera tells the audience that the production was made on a shoestring and to expect very little of the program. (Sparse sets can and do work, but they must be shot very tight, preferably using telephoto shots in place of wide angles, to collapse the empty space.) A set that appears to be fully developed sends a very positive message to the viewer. Development may, however, be accomplished by painted flats, by bookshelves (which can be rented), by colored geometric shapes, none of which costs very much. A good lighting director can add color and emphasis to certain set elements, and a good director can focus audience attention toward these pieces by careful placement of talent and strategic camera shots.

4. Hiring the Designer

Although the networks and some local stations employ their own designers (in local stations, the graphic arts department may double as the scenic department), most designers are free-lancers. New York, Los Angeles, Chicago, Toronto, and other large markets support a free-lance community that is available for commercials, motion pictures (frequently in LA, less so in other cities), industrial stage shows, and, of course, television. Free-lancers get their work by referrals, and set their own rates in consultation with the producers. Television designers are specialists, knowledgeable about lightweight construction, easy set and strike (putting up and taking down), use of colors that look good (and look good together) on camera, the effects of lenses and unusual camera angles, and so forth. Theatrical and motion picture designers can adapt to television with relative ease (some more easily than others), but designers who work, for example, in store display, are often beset with too many problems due to differing requirements to work efficiently in the new medium, and the set suffers (or the budget suffers) as a result.

Because the designer is usually called in on a specific project, the work is not speculative (except in some cases where several designers are up for the same job and portfolios do not provide relevant information). Design is a reputation business, where artists traditionally are paid for their first sketches, whether they are used or not. The deal is usually set up in steps, where the designer is paid for initial renderings, revisions, plan, and supervision of construction, painting, load-in, and sometimes maintenance. Most designers stick with the project from beginning to end; there is rarely time on a production schedule to replace a designer and start over.

GRAPHIC ARTISTS

Most television stations have an art department that is responsible for the creation of logos, artcards, slides, and just about everything else that's needed. Producers working outside the local television community generally develop lists of free-lance artists and independent graphic design companies.

Working with an artist is very much like working with a designer. The process begins with a concept meeting, is followed by a meeting to review pencil sketches, and is completed with the submission of graphics on schedule.

Some artists specialize in television work. Artists who are not trained for television traditionally make three errors in their early work, and each is easily corrected by reviewing work on camera prior to use. First, artists tend to underestimate the importance of the 3:4 "aspect ratio" of the television screen. Second, they tend to use too much detail, too many elements. Good television graphics are *simple* in both design and execution, and they are always designed to be seen through that 3:4 window. Lines should be fairly bold; thin lines may not read at all (or may appear to vibrate) on camera. Third, newcomers forget

to think in monochrome. Most television sets in America are poorly adjusted, and there are still a lot of black-and-white sets out there. Thinking in terms of contrasts and brightness ratios is more important than thinking in terms of color, as one would do in print. Blues, greens, and autumn golds work very well on camera; reds, oranges, and extreme whites or blacks can be troublesome, especially when they contrast dramatically with the background color. A video engineer will supply the range of "video gray tones," which lop off the blackest blacks and the whitest whites, which cannot be reproduced effectively with current video technology.

All of these rules are made to be broken by a clever artist. Some time spent in a studio can improve a television artist's work considerably.

ANIMATORS

With the advent of computer animation and related computer graphics tools in recent years, the use of animated graphics has become quite common in programs and in commercials. The networks routinely use animation in their promos, even on their news programs. Animation in network sports programming is generally state-of-the-art, particularly on NFL games, major league baseball games, and, of course, the World Series and the Super Bowl.

There are about a dozen top-notch computer animation houses that create, design, and produce animation for the networks, for sponsors, for anyone with a budget large enough to absorb the many thousands of dollars per second required for the best work. Larger companies dominate, but there are smaller animation houses in most major markets that can create impressive work on smaller budgets.

In any case, animation is a costly procedure,

generally used for a few seconds of an open, or a logo, and little more. The process begins with a meeting between designer and producer, who discuss specific requirements (content, design, schedule, budget). After a week or so, the designer returns with a storyboard and a final budget. The boards are scrutinized, revisions are made, and the designer starts working with the animator to create the piece. If it is complicated, a test animation will be made by substituting pencil sketches for completed drawings. In most cases, the animation is completed based on the storyboard.

Nearly all animation is now done with the help of a computer system, because it allows the animation team to skip over steps once done (rather tediously) by hand. In the near future, computer animation technology will blend with paint box technology, allowing producers to create longer sequences at lower prices.

MUSICIANS AND MUSIC LIBRARIES

Almost without exception, every television production should have some sort of music. Television music falls into four basic categories: opening (closing) themes, background music, bumpers (leading into or out of a commercial or break), and speciality music. Listen carefully to any well-produced television show, and you'll probably hear a dozen or more pieces, usually produced with some thematic unity. When preparing the list of music requirements for a new production, each application should be listed, with information about the pace, the use (e.g., as background music or as a theme), and, most important of all, the allotted time.

Existing music can be used for a production provided that all of the necessary permissions are granted. However, since every production has its own specific needs, television music is most often custom-produced.

1. Working with Existing Music

There are millions of songs available on records, and, in theory, the material exists to suit the needs of almost any production. Finding recorded material that is appropriate, within a reasonable period of time, is challenging at best.

Several libraries of recorded television (and radio) theme music are generally available at local stations. In some cases, the purchase price of the library includes all rights fees (see below). In other cases, an annual fee must be paid to the copyright owners for each year's use of the library material. Library music is perfect for some applications (local commercials, for example), but the selections are limited, and, despite the best efforts of the producers of the music, the pieces tend to sound generic. Sometimes, the theme that one producer selects from the library will also have been chosen by another producer. The advantage in library music is ease of use: the tracks are clearly labeled for their applications (e.g., "Opening news theme #3, fast and energetic"), and there are no clearances or permissions required beyond the contractual relationship signed at the time of the purchase, and, of course, the renewals.

It is possible, in theory, to use any recording on a local television broadcast. Stations pay annual fees to ASCAP and BMI for blanket clearances of compositions (which takes care of the publisher of the music and composer). Separate fees should be negotiated with the artist, who is not covered in this blanket clearance, and with the record company, who is likely to be a copyright holder as well. The only problem with this formula is the assumption that the artist and the record company will negotiate; the amount of money involved does not justify the administrative time and effort, so it is easier for the record company or the artist to say no, unless the production will be seen by a large-enough audience to dramatically affect record sales.

This becomes even more complicated when a recording is to be used on a home videocassette, or on a corporate video presentation, where the music is part of a producer's effort to make money. In these situations, everyone may feel that they are entitled to a small piece of the action (which means negotiating separate deals with the composer and publisher, the artist, the record company, and sometimes the producer of the track). The administrative headaches, which require semiannual updating for the life of the production, are enough to dissuade most producers from using existing tracks. On the other hand, it is possible to make the arrangements, and in some cases, the inclusion of popular music will greatly enhance the impact or the value of the production. In Los Angeles and in New York, there are companies that specialize in music clearances to relieve producers of the time-consuming negotiation and follow-up paperwork.

Synchronization rights ("sync rights") are granted as permission to use a particular composition; mechanical rights are granted as permission to use a particular recording. Both rights must be cleared before a recording is used.

There are many gray areas in the use of recorded music on television, so it is wise to either work within existing guidelines or to consult the business or legal people for policy decisions. Music video, which may be considered promotional material as opposed to program material, is one such area that may require legal interpretation.

A list of every piece of music, its running time, artist, publisher, and record label, called a music log, should be packed with each completed master tape. A copy of the log should be placed in the show file as well.

One final point: there is no official policing body that inspects every television production for proper music clearances (although spot checks are common if a producer or production company is suspect). If a production is broadcast or cablecast, and is likely to be seen by many people, clearance rules must be observed absolutely. If the production is being prepared for internal use (e.g., within a company), and only a limited number of people will see the program, the clearance work may not be worth the time or trouble.

2. Creating Original Music

Since the process of clearing recorded music can be so complicated, and because few pieces of recorded music match the needs of a production precisely, much of the music heard on television is custom-made. Creating music for a prime time series is very much like scoring a motion picture; it is a postproduction process that usually begins only after the composer has seen a rough cut (sometimes, however, a composer starts work as soon as a script is available). Creating an opening, a closing, and a few bumpers is much more like recording a song; working with rough timings, the composer and music producer work together with a small ensemble and record the necessary pieces with a minimum of fuss. With certain program formats now standardized, there are now companies that sell talk show music, news packages, and promotional music packages to one station in each market on an exclusive basis. The needs of WCBS-TV news in New York may match the needs of WBBM in Chicago and WCAU in Philadelphia, and if the needs are identical, there is no reason why each of these CBS stations must have its own music. (This packaging concept has been common in radio for many years; it is more a recent development in television.)

A. Working with the Composer

Once the producer has made a basic list of music requirements, he calls a meeting with the composer to flesh out the ideas, to develop some basic musical concepts, and to set the schedule and budget. If the music is to be recorded before the studio production (typical in nondramatic programs), the composer is given a list of pieces to be created. If the music will score a dramatic presentation, the composer is supplied with a rough cut of the program, a script, and some thoughts from the producer or director about sequences that seem to require music.

The composer leaves with instructions, and returns, usually a week or two later, with the major composition work complete. It is typically played on a piano, or, if no piano is available, it is played from a recorded cassette. Some revisions are made, and the composer completes the score.

An arranger then takes the basic composition and writes charts for each of the musicians in the ensemble. (In many cases, the composer and the arranger are the same person.) Copies of the charts are made by a trained copyist (a union job).

If the composition is an original, it is published by either the music producer, the composer, the television producer, or the production company (depending upon the deal, which may have later implications, if, for example, there is a sound track album).

B. In the Studio

In television, the composer or arranger usually serves as the producer of the music sessions, selecting the musicians, booking the recording studio, making arrangements with the recording engineer. If arrangements are to be made with the AFM (the musicians' union— see below), the producer takes charge of these as well.

Working closely with the television producer or director, the music producer supervises the laying down of each track (there are usually sixteen or twenty-four tracks, sometimes more), and then works closely with the recording engineer to set up a mix. In most cases, the rehearsals, the laydown, and the mix

are done over one day (frequently one long day).

Multitrack tapes are carefully labeled and stored, and the resulting mixed master, typically a ¼-inch audiotape, is delivered to the television producer.

3. The American Federation of Musicians

Unlike most of the other television unions and guilds, the AFM (or "AF of M") has chapters in most small, medium, and large cities. They are not exclusively a television union; their contracts cover live events, concert performances, concerts, motion pictures, radio, records, just about every area of entertainment in America (and in Canada). The AFM has rates and rules that cover everything from the composition and rehearsal pianists to copies and arrangers. If one musician "doubles" (plays a second instrument), the producer is charged. If a bassist or other player of a large instrument is required to transport his or her own instrument, the producer is charged for carting. If the music is recorded in a studio and later replayed on a sequence of productions, the musicians must be paid as if they were in attendance at each of the productions. The AFM rules are comprehensive, all inclusive, and, in many situations, very carefully policed. It is best to work out the initial budget for recording original music by working closely with an AFM representative, or with somebody who knows their rules well. There are many charges listed in the contracts (which are available to any producer for a few dollars), and some of them are obscured in the complexities of the rates and rules and charts and residuals.

Most professional musicians carry AFM cards, and it is very nearly impossible to assemble a group of capable players in any American city without involving the AFM.

Nonunion workers, though they do exist, are not as common in music as they are in other disciplines.

Like most unions, the AFM surcharges all payments to members with a contribution to their pension and welfare fund.

TEN

Business Affairs, Program Distribution, and Financing

Business affairs specialists deal with contracts, distribution, and financing.

Although the business aspects of television can be quite complicated, most producers and directors are exposed to only four areas: (1) negotiating their own employment agreements, (2) negotiating agreements with their employees, (3) rights agreements, and (4) distribution agreements.

THE PLAYERS

1. The Business Affairs Department

Most of the larger television companies employ one or more attorneys in the business affairs department. These are the people who negotiate and draft agreements, who protect the producer, packager, distributor, or network from both business and legal problems. (A small percentage of business affairs people are not lawyers, but come from management, finance, or other related areas.)

In a small production company, employing between six and twelve people, there is likely to be a head of development, a head of production, and a head of business affairs. In a larger company, there might be a VP, Business Affairs, with several attorneys and assistants on staff. Very small companies take care of some of their own business affairs and hire a lawyer to do the rest.

2. Outside Counsel

Because entertainment law is a specialized domain, it is wise to work with an attorney who has had considerable experience in television, music, motion pictures, theater, or a similar area. There are many distinguished firms in Los Angeles and New York with a particular expertise in the entertainment industry and some firms in Boston, Washington, Toronto, and Chicago with one or more specialists. In other cities, there is little call for an attorney with specialized entertainment experience.

The range of services required of an entertainment lawyer varies considerably. Some producers use lawyers to negotiate agreements, while others prefer to negotiate on their own behalf, consulting with the attorney only in the final stages or with regard to problem areas. Other producers have agents negotiate (see below) who may have house attorneys to review agreements. A producer with an entrepreneurial bent usually needs a lawyer to draft agreements from scratch.

When working with outside counsel, it is wise to establish the rules before a project begins. Decide who will negotiate, which side will draft the agreement, to whom the agreements should be sent, who will communicate with the other side and on what subjects. To save money, experienced producers separate "business points" (e.g., how much will be paid, royalty percentages) from "legal points" (e.g., the inclusion of a *force majeure* clause, which permits a party to disregard the contract in case of strikes, fires, lockouts, and so forth). The producer handles the business points and then communicates the information to the lawyer, who takes care of the legal points, pending approval by the producer, of course.

Lawyers charge by the hour. Fees vary based on the size and standing of the firm, experience, and marketplace conditions (and, sometimes, personal relationships).

3. Agents and Managers

Creative people, as a rule, are poor negotiators. The reasons can be complicated, but they all come down to a creative person's overwhelming desire to do a project, and, in some cases, to be well liked from start to finish. Producers, directors, writers, performers, even designers and composers, can hire representation for a fixed percentage of earnings (usually 10 percent, sometimes 15 percent). Agents do more than negotiate; if they're good, they can help the creative person to find work, or to sell projects.

The two best-known agencies are the William Morris Agency and ICM. Each employs agents in New York, Los Angeles, London, and other cities to represent clients; individual departments handle clients for television, music, motion pictures, personal appearances, and publishing. Other agencies, some very good, some very bad, most somewhere in between, handle clients who either prefer a smaller company or are not yet successful enough for a larger agency. As one becomes established, particularly in New York or LA, representation is generally available when it's needed (usually to negotiate an agreement or to help select among offers, rarely to "find work" early in a relationship).

Is there a difference between an agent and a manager? There is for a performer, who looks to the agent to find work, field offers, and negotiate, and to the manager for personal career guidance (e.g., developing an act). The difference is less clear for producers and directors, whose careers tend to need less outside management.

If you have an agent, do you need a lawyer? Agents negotiate the best possible deals for their clients, but few of them are really familiar with all the legal minutiae that appear in a contract. Even if the agency has its own legal department, a lawyer should read any agreement before it's signed, just to be sure.

4. Representing Yourself

There's a saying, "A doctor who has himself for a patient is a fool." In most cases, the same thinking can be applied to creative people who want to negotiate on their own behalf. If you do decide to represent yourself, make certain that you are aware of current marketplace salaries and fees, and try your best to research similar deals. Whether the negotiation concerns a small company producing its first industrial, or your employment as the director of the weekend news, handle all aspects of the discussion in a friendly, agreeable, and businesslike manner. Concentrate on the main points: the fee and the term of service. Hang tough when you feel strongly about a point (but be careful of using terms like "deal breaker," unless you really mean to break the deal over a single issue). Make a list of every point you wish to cover, but plan to work through this list only once (people get tired of running through list after list, particularly when a negotiation should be fairly simple, or when it is nearly done). Learn to compromise, but never "give something away," particularly something that could come back to haunt you later on. And, if it is your style, feel free to do some trading, giving in on certain points in exchange for having it your way on others. Avoid discussing small points extensively; when a small issue becomes troublesome, leave it for the moment and return later, after all the key points are resolved. Above all, make certain that you can live with the terms of the agreement, that you understand the terms completely before you sign, and that there is no question of your responsibilities (many producers and directors have been shocked to find that a fairly general interpretation of an agreement allows the employer to take them off one show and require them to work on another, for example). Find a good attorney, one who specializes in entertainment, and listen carefully to what he or she has to say.

When you enter into a negotiation, be prepared to make firm commitments, but try to leave yourself an out if you run into trouble. If you're negotiating on behalf of a company, don't feel as though you must resolve every issue alone—discuss the more troublesome issues with partners or associates. (Herein lies a benefit in having someone else negotiate for you—he or she can never be pinned down without "checking with my client." If you're out there alone, you may have no such option.)

Above all, when you make a commitment, stick to it. Don't go back on your word unless you absolutely must, because this severely hurts your credibility both as a negotiator and as a new employee.

Once the negotiation is complete, do as the agents do: write a brief memorandum to the other party covering the key points, keep it simple, make no attempt to do anything more than list the major points in the least complicated way (anything more could conceivably be viewed as a legal contract, which should be written only by a lawyer). And, by all means, make certain that the other party understands the deal the way you do. In most cases, a brief legal document between the two parties is appropriate.

CONTRACTS

A contract is a written agreement between two or more parties that details the understanding between them. A contract may be quite brief (one or two pages long) or upwards of fifty pages for complicated agreements. Usually it is ten to twenty pages long. Contracts are usually drafted by attorneys, who work with a combination of "boilerplate" language (e.g., standard paragraphs, usually stored on a word processing library disk and simply run in the proper places in each new agreement) and language specific to the deal as outlined by their client and set forth in discussions with the other party or parties. Once drafted, the agreement is reviewed by the client, who usually suggests changes. A revised agreement is then sent to the signatories for their review. The order in which a contract is signed is a matter of personal style. Ideally, all parties are present to sign an agreement. In real life, the contract is usually signed by one party, then sent to the other(s) for signing. It is only when all signatures are in place that a contract is legally binding.

Although they may be worded informally, all letter agreements and employment agreements should be considered to be contracts, and should be reviewed by an attorney before signing.

1. Employment Agreements

For most producers and directors, contracts and employment agreements are synonymous. An employer asks an employee to sign a contract to guarantee services over a specified period of time, at a specified rate. An employee who signs a contract feels a sense of security, for the contract guarantees a flow of income. Bearing in mind that employment agreements are generally written for the benefit of the employer, here are the basic facts that should be included in an employment agreement:

A. *The term of employment*, including a start date, the number of days/hours to be worked each week, and/or a guaranteed minimum number of days/weeks/months of employment. Every situation is different; just be sure that the understanding of employer and employee match.

B. *The job responsibilities*, described as specifically as possible.

C. *The salary or fee*, and how it will be paid. In most cases, a producer or director is paid weekly or biweekly along with the other company employees. When productions are being taped ahead (e.g., thirteen weeks of shows produced and recorded over the course of ten weeks), the producer may be paid either based on the ten weeks of actual production work or on the thirteen weeks of results. This situation must be negotiated in every case (and it tends to become important if a producer, director, or performer is terminated earlier than expected).

In some cases, usually on project work,

payment is due on successful completion of elements on the production schedule. On a business video, for example, the producer might be paid 25 percent to start, 25 percent upon acceptance of the shooting script, 25 percent upon completion of shooting, and 25 percent on delivery and acceptance of the master. There are variations on this schedule, where larger amounts are paid up front (the employee's advantage, usually the result of a good negotiation), or held until certain checkpoints are cleared (the employer's advantage, used when the employer is unsure of the results, or when the employer is tied to a similar contract of his/her own).

D. *Additional compensation* for overtime, work done on weekends and holidays, work done out of town, and so forth. All of this is negotiable. Some has already been negotiated in union agreements.

E. *Health insurance and benefits* packages, pension plans, profit-sharing plans (e.g., if the show is a hit, and it makes a lot of money, the producer or director should receive some financial reward, be it profit-sharing or a bonus of some kind), and, if applicable, stock option and purchase plans. Also: the number of vacation days, personal and sick days, company holidays and so on.

F. *Negotiated points* about the size of the staff and control over them (including the right to hire and fire with or without approval from superiors); control over the format, the budget, the schedule; smaller details, from the size of office to title and screen credit.

G. If the employee is going to be *developing projects* while under this agreement, the parameters should be established during the initial negotiations (while you still have some power—many projects have been taken by employers because creative employees were unwilling to leave the company while bargaining for equity). Discuss participation in the profits (and watch out for clever wording in the agreement; have an experienced attorney define terms like "net collected revenues" in detail), ownership, copyright, future involvement if the project is sold to a third party (a syndicator, for example), and, perhaps most important of all, the fate of projects developed by the employee if the contract is broken, or if it expires. Any properties owned by the producer or director prior to the agreement should be listed, with a clear understanding of their fate once the association is over. In a business where so many creative people become involved in germinal projects outside of their basic employment, the term "conflict of interest" should be discussed as well. Agreements involving project development tend to be complicated; they should be negotiated by an agent or lawyer in almost all cases.

H. *Employment exclusivity* should be carefully detailed as well. If the employee plans to moonlight—many directors do—it is wise to do so without threatening a job. Some employers will permit nonexclusive arrangements if they are noncompetitive. In other words, a local news director for WNEV would not be permitted to fill in across town at Boston's WCVB, but he or she might be allowed to direct a documentary for WGBH, the public television station, if time permitted.

When working on a project basis, particularly if the employee possesses a special talent or expertise, many employers require some limitation of future services rendered to competitors. In other words, they do not want the producer to use what he or she has learned to create a competitive product. These situations always require careful negotiation, because the producer's ability to earn a living may be dependent upon the flexibility of the language in this section of the agreement.

I. The terms of *severance* from the employer. If fired, how many weeks (days) of pay will be due? How long will the benefits package run? Can you pay for an extension? If the employee decides to leave, how many weeks' notice are required? What is forfeited by leaving without notice?

J. Bearing in mind that contracts are written for worst case scenarios, *consider every possibility*. What if the employee is injured and cannot perform? Or pregnant—what are the terms of the maternity leave? Or paternity leave? What if the employee is hired as a director—and is not permitted to direct? Or hired as a producer, only to find that the previous producer decided not to leave after all? Be reasonable, but be firm if you really believe that a clause should be included in the agreement.

2. Employee Agreements

Once hired, the producer, or executive producer, will be expected to put together a staff and to negotiate deals with each of them. In most cases, the negotiations are limited to salary, benefits, and title, and a handshake is all that is required. When a contract is necessary, as it will often be when hiring a director, writer, or performer, first find out whether standard agreements exist (there are standard DGA, WGA, AFTRA, and SAG agreements, for example). If they do not, ask the lawyer or business affairs person to draft a short letter agreement, listing all the key points in two to five pages. If an associate producer or PA asks for a letter, keep it to a page listing key points and nothing more. This kind of thing will usually suffice:

Name: Sharon Hasenauer

Social Security Number: 088-88-8888

Job: Production assistant

Fee: $200 per week, for three weeks

Start Date: September 11, 1986

Benefits: None

Expenses: Reimbursement within two weeks after submission of receipts expense account form; prior approval required for all expenses over $10

Be sure to include address and phone number for future reference. Both parties should sign the agreement, so that it will be binding.

3. Agreements with Independent Contractors

Until this point, the party doing the hiring has been called the "employer," and the party doing the work has been called the "employee." Increasingly, creative people have been forming their own corporations (sometimes with only themselves as employees), and offering their services as independent contractors. When hiring a person through a corporation, there are no taxes withheld. The corporation is responsible for paying its own taxes. There are other benefits as well; a good accountant (preferably one who has clients in a creative field) can be helpful in this area.

4. Other Types of Contracts

Three other types of contracts are common in television production: the rights agreement,

the distribution agreement, and the co-production agreement. These agreements tend to be quite complicated; a good entertainment attorney can be very helpful in outlining the details of each of them. Briefly, here are the key points:

A. A *rights agreement* sets forth the arrangement between a producer and the owner of the rights to, for example, a book. In essence, the producer is buying the exclusive right to use the title and the contents of the book, for a negotiated price, and, perhaps, a percentage of the profits. Producers usually establish the purchase price and then an option price (usually a small percentage of the purchase price) to "tie up the rights" for a 60-, 90-, or 120-day period. If the option is exercised, the purchase price is paid. If it is not, the rights holder keeps the option money and the contract is forgotten. The rights agreement usually includes a considerable amount of legal language that assures that the rights holder does indeed own the rights, and so forth. Sometimes, the rights holder will require an active part in the production, or script approval, or approval over the final cut—all of this is negotiable. In fact, just about everything in a rights agreement is negotiable.

B. A *distribution agreement* is written between the person or company who owns the rights to a television production (e.g., "the producer") and the company who will distribute it. Typically, a distribution agreement will list the markets where the production can be sold (e.g., domestic home video, foreign home video, domestic syndication, foreign syndication, basic cable TV, pay cable TV, direct mail, and many other markets and subdivisions). The distributor usually pays a percentage of its collected revenues to the producer; when a project is valuable, large up-front guarantees and "advance" payments are common. There are many types of distribution arrangements: some base their payments on "gross collected revenues," others on "net collected revenues" (the definitions of these terms vary; the definitions are important because they determine how much money will be kept by the distributor for expenses related to the sale). Every distribution agreement should be carefully scrutinized by a knowledgeable attorney.

C. A *co-production agreement* usually binds two or more parties in the production and the ownership of a television project. The parties each contribute something of value, usually a sufficient amount of money to develop and/or produce the project, sometimes just the idea or the rights to a book or a character. The parties are required to fulfill certain obligations by certain dates specified in the agreement (e.g., 50 percent of the production budget upon the start of principal photography—the first day of shooting). In some cases, one of the co-producers (also called the "co-venturer") is a distributor, or a party interested in owning certain distribution rights. In other cases, the parties are financing for tax reasons. Once again, every situation is different, and, once again, every situation requires the close supervision of an experienced attorney.

PROGRAM DISTRIBUTION AND FINANCING

Money for television production generally comes from one of five sources:

1. A television network, like ABC, CBS, NBC, and, in some cases, PBS, or HBO or Showtime, or a station group like Westinghouse.
2. A sponsor, like Procter & Gamble, Mobil, Nabisco, or IBM (or, in the case of local broadcasting or cablecasting, a regional or local advertiser).
3. A distributor, such as a syndicator (Viacom, Worldvision, King World, etc.) or a home video label (Vestron Video, CBS/Fox, Embassy, etc.).
4. An investor, be it a studio (like Universal or Columbia Pictures), a foreign partner (like England's BBC or Canada's CBC), a company interested in owning television properties, or private individuals.
5. A company that simply needs a videotape produced to promote or explain a product or service, like a record company in the case of a music video, or an advertiser in the case of a commercial, or almost any type of company in the case of a business or training videotape.

There are many versions of "creative financing" for television production, combining two or more of the above with one another or, in some cases, with more arcane methods of raising money. When television was simply three networks and their local affiliates, financing arrangements were fairly standard. With the increasing amounts of activity in syndication, the various forms of cable and pay TV, home video, business video, and foreign participation, the rules are changing.

A music video, for example, is traditionally financed by a record company as part of the promotion budget for a single release. If the record is associated with a major artist, production financing may be forthcoming from pay and cable television in exchange for exclusive showings, and from home video as part of a package of videos made by the artist. The distinction between production financing, traditionally paid before the master tape is complete, and distribution proceeds, traditionally paid upon delivery of a master tape, becomes inconsequential when the distributor advances payments in exchange for rights. And this is becoming increasingly common.

A training tape provides a contrasting example. Here, a company pays for a production for internal use, providing complete financing. Enter a home video company with an interest in recutting the master for release to their market, or a local PBS affiliate interested in creating a network show, and the entire plan is revised to accommodate the new situation. Television production is expensive; it frequently pays to have partners.

How the Markets Work

1. Network Television (ABC, CBS, NBC)

In productions made for prime time, the networks usually pay for the right to air the show once, and the option to air the show a second time. The combination is sufficient to cover the cost of production on a typical half hour, hour, or movie-length presentation. If the network "picks up the option," and airs the show a second time, the owner of the show makes more money (and so do the director, the writer, and the performers, whose union contracts guarantee "residual payments"). After the program airs twice (or, in some cases, three times), rights revert back to the owner, who usually makes a deal with a syndicator to earn additional income (relatively easy to do with a movie, harder to do with a special, nearly impossible to do with a series unless it has completed a substantial number of episodes—or new episodes are subsidized by the sponsor).

Most prime time programs, including motion pictures, are produced and owned by the studios (Warner Brothers, Columbia Pictures, Universal, Twentieth Century-Fox, etc.) or by the large production companies (Lorimar-Telepictures, Aaron Spelling), frequently in association with a smaller production company. Individual producers and creators of these programs generally participate in the profits in syndication and other "aftermarkets." The networks also produce a limited number of prime time dramas and comedies, such as "Moonlighting," which is produced by Glenn Gordon Caron's Picturemaker Productions in association with a division of ABC (known as ABC Circle Films).

The arrangement between network and owner is similar in daytime programming and in sports programming, but these usually air only once, so the deal is altered accordingly.

News and documentary programming is generally produced by the networks, not by outside companies.

2. PBS

There are several sources of financing for programs seen on PBS. The network maintains a budget for productions and acquisitions (completed productions, "acquired" for one or more airings), as do some of the larger stations (like WGBH). Whenever possible, public television looks to the Corporation for Public Broadcasting (or "CPB," which offers money from the government and from other sources), foundations (like the Ford Foundation) for grants of production monies, and to large corporations (like Mobil and Exxon) for the same. Because financing arrangements can be complete or partial, and made with either the network or local affiliate, there is a great variety of production financing and distribution arrangements related to programs seen on public television.

3. Syndication

The majority of syndicated programs are "off-network" properties, as described previously. A syndicator offers a percentage of the profits to the owner of the program in exchange for distribution rights (usually with some sort of guarantee, which is large for top properties and small for those that are harder to sell).

Syndicators are also very active in providing original programming to local stations. Game shows like "Wheel of Fortune" and "Jeopardy!" have been enormously profitable. Here, the syndicator pays the owner a fee that covers the production budget and, one would assume, a substantial profit. The syndicator makes its money by selling the program to one local station in each market (or in as many local markets as possible), for as much as possible. At least seven or eight of the top ten markets are usually necessary for a syndicated program to be successful, with 100 of the 150 largest markets needed as well (since the fees charged to the stations vary with each syndicated show, these numbers are different every time). Most syndicated properties, particularly those created specifically for this marketplace, are not likely to clear a sufficient number of stations, so many pilot programs are produced each year to increase each syndicator's odds. In order to limit the risk, syndicators may share the costs of production with the producer/packager, with investors, or with a sponsor in a "barter" relationship.

In "barter," the syndicator and the sponsor provide the program at no charge to the local station. The syndicator charges a fee for services, and the sponsor provides the program with, for example, four of the six commercial slots filled with its own advertisements, sometimes with a cash incentive paid to the station as well. If the station can sell the other two, it makes a profit. (Sometimes the station does well with this arrangement; most barter programs are not likely to generate huge ratings, however, so stations use them mainly for filler, late at night, on weekends, etc.) In a barter arrangement, the production is financed by the sponsor, and, as might be expected, profits are usually slim. When the system works, however, the sponsor, the stations, the producer, and the syndicator all benefit.

Station group owners will, on occasion, use the syndication mechanism to create a "fourth network," as with Fox Television and Operation Prime Time. Here, the production is financed either by the station groups, or, more often, by a combination of the groups and the studio, which subsequently owns syndication rights.

4. Local Stations

Local stations finance their own news, sports, and public affairs programs, and produce them as well. Original programs on local stations are generally created in response to sponsor interest (or, in some cases, an important news story). In these situations, the sponsors usually pay for most of the cost of production.

Given the choice of picking up the tab for a special, or simply buying a few spots in the news or in a syndicated show, most sponsors elect the latter. Cost is the principal reason why few local stations produce very much local programming. (If a station has excess production capability, such as free studio hours in between the morning show and the evening news, this studio time is usually sold by the hour for the production of commercials—this is far more profitable than program production, and far less risky as well.)

5. Pay TV

As a rule, the pay TV networks, HBO and Showtime, finance their own productions and, in most cases, maintain ownership as well. Producer/packagers are generally paid enough money to make the program(s), plus a small profit. The networks have the right to play the program many times over a two- or three-year period (their schedules demand a high number of repeat plays), and to renew for a longer period at their option. If the network owns the show—and this is the case in the majority of the situations—it may decide to make a home video deal, sell it in the foreign markets, and so forth. If the network does not own the show, but pays a license fee to the owner for a specified number of plays over a specified period of time (a "license deal"), it may provide partial financing in exchange for partial ownership or a share of the profits as well. Once again, there are many types of deals.

6. Advertiser-Supported Cable Networks

There are a handful of advertiser-supported cable networks, like Nickelodeon, MTV, and WTBS/The Superstation, which produce original programming. Money for these productions comes from a programming budget, which gets larger every year as these networks grow, or, more often, from sponsors.

As the "penetration" of these networks increases ("penetration" is the percentage of U.S. television households that can receive the networks), their desirability to advertisers increases as well. MTV, even in its early years, has been active in arranging co-productions with sponsors, with home video companies, with foreign broadcasters, and with syndicators. This flexibility has resulted in a wide variety of excellent programs for the network (including their annual awards show, which is first seen on MTV and then syndicated to commercial broadcast television stations at a later date), at relatively little cost to the network. These deals are not necessarily originated by the network; producers, advertisers, and syndicators have all picked up on the network's interest in special distribution arrangements.

7. Local Cable Systems

There isn't much money here, because the number of viewers who see any one cable system is relatively low (a few hundred thousand subscribers at best, and only a tiny percentage of them actually watch a given program on a given channel, unless it is supported by national promotion). Programs produced for local cable TV are either financed by a producer and his/her friends, by local sponsors, or, in some cases, by a local system, school district, or municipal organization. Some of the larger cable television systems, like Long Island's Cablevision, have experimented with local news programming.

Such situations are extremely useful for those interested in learning the television production business, but it is exceedingly difficult to produce programs that make money in cable television (as always, there are exceptions, but they are few and hard-earned).

8. Home Video

Despite the considerable media attention paid to original home video programs, the home video companies make their money by selling movies on videocassette and tend to treat originals as a sideline. Most home video companies are very selective about titles that they produce (e.g., finance and supervise), or co-produce (e.g., finance with a partner, like an investor or a pay TV company). The reason is in the numbers: it is easier to sell twenty-five thousand or fifty thousand copies of a motion picture title, even one that met with mediocre success in the theater, than it is to sell an original title made for home video. Increasingly, originals with celebrities, licensed characters, or some other "presold" commodity are finding success alongside the movies in the home video marketplace.

In a relatively short time, home video has become an active consumer products business, which operates in two parallel ways. In the rental scenario, videocassettes are sold by the labels (or their distributors) to the video rental stores, who rent the products to make their money. None of the rental income finds its way back to the label; the store keeps all of the money. In the sales scenario, videocassettes are sold by the labels (or their distributors), and then resold by the stores with a markup (the stores buy at wholesale prices and sell at retail prices).

Motion pictures tend to do well in both the rental and the sales scenario, especially if they were successful in the theaters (although certain "box office bombs" do become hits in home video). Nontheatrical product, such as original children's programming or exercise shows, tends to do best as sales items. Some videocassettes are sold in bookstores, record stores, and specialty outlets, because the video labels recognize that the video stores are best at selling or renting movies. Nonmovie product may do better elsewhere (e.g., health-related titles in drugstores, music videos in record stores, videocassette game titles, such as "Clue," in toy stores).

The home video labels generally pay the cost of production in exchange for exclusive, or semiexclusive, rights, and offer a royalty as well.

9. Business Video
With higher prices (as much as $500 per product), and lesser production requirements, many business video products have been very successful. Business videotapes are sold to corporations, usually for executive or management training (on topics like running successful meetings, closing sales, and developing team spirit). A typical product includes several shorter tapes (nobody likes to take two hours out of the workday to watch television), workbooks, leader's guides, and other learning aids. Production values are usually minimal; there is no reason to create lavish stage settings or flashy special-effects sequences for most situations.

These productions are financed by publishers of business information, by distributors who concentrate on this market, or by independent producers who wish to maintain ownership, selling their wares via direct mail and at conventions. This business has "hits" like any other—and the right title can make a lot of money.

In-house videotapes generally are financed by the company to fit a specific need. There is rarely an "aftermarket" for these products.

10. Educational Video
Through the years, many hours of educational television have been financed by school systems, colleges and universities, government organizations, corporations, publishers, and individuals. Some of these programs have been quite successful, seen in hundreds, even thousands, of schools, on local public television outlets, and in a great many special situations. Rarely do these productions earn enough money to cover their production costs; they are usually produced to teach a specific subject, and their real value is in their effectiveness, not in their profit potential. This situation is not expected to change in the foreseeable future (but, once again, there are some very exciting exceptions).

11. Commercials
Television commercials are financed by advertisers, who usually pay a fee to an advertising agency who produces the commercial campaign, and an additional fee for the placement of those commercials on the air.

12. Music Video
Music videos are promotional tools, used by record companies to help sell products. Most are financed by the record companies, who pay a producer or production company to create the video within a specified budget. Understanding the importance of these videos, many artists and managers choose to supplement the record company budget with money of their own. In most cases, the videos are owned by the record companies, but some artist contracts allow either shared ownership, or, in the case of some major stars, ownership of the videos by the artist. Based on the success of "The Making of Michael Jackson's 'Thriller,' " and some of the more popular compilations for home video, it is easy to see the value in owning the rights to these products.

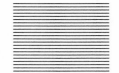

ELEVEN

The Real World

To many people, even readers of this book, television is a magical profession where creative people spend a great deal of other people's money to fulfill their own visions of "what the people out there should be watching." An inside look, however, provides a realistic picture of everyday life in the *business* of television production, where many decisions are made by the wrong people, for the wrong reasons, where the giant television machine resists new ideas and innovation, where tremendous amounts of money are spent foolishly, and where small amounts of money required for necessary improvements can be hard to get. In other words, television is no different from any other business; it is not perfect. Still, there are benefits. It is a glamorous business because of the celebrities (remarkably few producers or directors actually work with the superstars, but most outsiders think they do), and because so many people see the final product. And because it is a communications business, television seems to be blessed with an inordinately high number of people who are smart, clever, bright, outgoing, and funny (sometimes to a fault). It is fun to work in television, for even when things go badly,

most people who work here can hardly think of doing anything else to earn a living.

LEADERSHIP

The principal quality evident in successful producers and directors is an ability to lead people, to manage, to provide the enthusiasm and guidance that encourage a singular vision and a singular purpose: to create the best possible production, on time and on budget. A great deal has been said, and has been written, about the qualities of leadership, and the best ways to improve one's leadership skills. For some people, these can be useful. For most people who work in television, however, a combination of instinct and experience plays the largest part. This is partially because the ability to lead is directly tied to the ability to make the right creative decisions, and partially because people who work in television tend to adapt working styles and attitudes that more closely resemble those of artists, writers, and musicians than those of middle managers. Most producers and directors learn by observation, frequently working with a mentor, and then try to make their inevitable early mistakes on productions where the stakes are low.

What makes a producer or a director a good leader? Usually, it's a combination of elements. It starts with respect, which comes from making the right creative choices, providing credit where it is due, offering criticism without creating an adversary situation, making everyone feel a part of the team. Choosing the right people is important as well; a producer is always judged by his or her team, and their positive feelings for the project. If they're hard to work with, the producer will have a tough time getting everyone to work together. Motivation is essential, too; the producer and director must care more about the production than about anything else. Creating any production is a complicated task; it requires both the involvement and the interest of everyone in the cast, on the staff, and on the crew. If a producer or director doesn't seem to care, nobody else will either.

There is also the issue of money, and the way that it is spent. Producers who are either too conservative or too liberal with the cash tend to lose the respect of their co-workers. Directors who waste valuable studio time or editing time by doing work that should have been done beforehand, or by unfairly insisting upon a new setup or retake, will lose some respect as well. Whether they say it or not, everyone is aware that television production costs a lot of money, and that extra hours can put a project over budget. A producer who "cries poor" rarely has much credibility; he or she should never have allowed the production to get into that position. Worse yet, some producers and directors make the mistake of unintentionally creating a "class system" where money is concerned. On a location shoot, for example, providing the above-the-line personnel with their own hotel rooms and asking crew members to double up demonstrates a foolish and insensitive attitude. Offering low

per diems to the crew (a "per diem" is a daily cash allowance for meals and personal expenses away from home) while taking the star out to the best restaurant in town also shows a lack of consideration.

Now there's nothing wrong with managing the budget, and the entire production, according to one's own style and unique requirements. Everybody has his own way of doing things, and, as long as the show gets ratings, or the company president or client is pleased, this is what really matters. But every production is easier to make when it is guided by a capable leader. And, more often than not, the results show in the finished product.

BEYOND THE GLAMOUR

It's hard to be a model leader all the time. There are days when even the most patient of producers wants to muzzle the writer, to send the director home for acting like a child, or to call the whole thing off and start fresh tomorrow. Fortunately, these situations are not very common.

Producers, associate producers, and talent coordinators who work on network, syndicated, and major market talk, news, and variety shows receive a dozen or more free, newly published books every day, are invited to screenings of all the new movies, previews, and parties, usually with opportunities to meet celebrities, usually several times a week. An active producer, particularly one working in network television, can easily eat breakfasts, lunches, and dinners at the best restaurants in town on someone else's expense account. It sounds like a glamorous life, but even this routine can become wearisome.

The books, for example, are sent for review. Most are discarded or given to the secretaries; the ones that could provide the basis for show segments, however, must be read, or at least browsed thoroughly. This is a standard routine, followed week after week, usually in the evenings, after working hours. When an author is coming to town, someone (rarely the interviewer; usually an associate producer) must read the book, write a background summary, and create a series of interview questions for use on the air.

The free tickets to shows, the invitations to openings, and so forth are sent with the understanding that someone with the program will attend, at least in the case of the more important events. In New York, Los Angeles, Chicago, and Washington, D.C., there are several events every night of the week. One producer, whose job requires frequent attendance at comedy clubs where the big acts don't go onstage until 10:00 or 11:00 P.M. is out at a club, a screening, or a party just about every night of the week. This demands a special kind of commitment, and, of course, personal and family relationships that are extremely flexible.

Producers and directors who specialize in early morning programs, commercials, and in location shoots are accustomed to a very different schedule. For them, the workday may start at or before sunrise, and, in the case of outdoor location shoots, it is likely to continue until dark (and, in the case of the production staff, who must prepare for the next day's shooting, or the director, who usually reviews the day's results, or to plan the edit, the day may continue until 8:00 or 9:00 P.M.). One must be flexible in these situations as well; social engagements, theater and concert tickets, family obligations, any personal life at all must come second to the production's schedule.

MAKING A LIVING

Despite the glamour, and perhaps even because of it, most television salaries are not particularly high. People who work in shipping departments and as file clerks usually make more money than production assistants, and work shorter hours as well. Most producers and directors working in local stations outside the major markets, for small advertising agencies, and in business video are paid a livable middle manager's salary, sometimes a little more, frequently a little less.

Producers and directors who work in larger markets, at senior levels for large corporations or advertising agencies, and in the entertainment business earn above-average salaries. The producer or director of a prime time network show, for example, may earn $15,000, $20,000, or more per episode (at fifteen or twenty episodes per year, in 1987 dollars, this is a very substantial salary). On most network shows, in fact, the salaries for most senior staff members are considerably higher than those paid their local TV counterparts, because the talents are so very specialized, and the investment usually pays off in good ratings.

This is the case in major market local news as well, where the best producers, directors, anchors, and reporters may earn up to $100,000 per year and more (1987 dollars). The reason is high ratings; a producer, director, or reporter who has worked his or her way up from smaller markets is usually hired on talent, reputation, and the ability to maintain a share of the audience that's higher than the competition (which in turn generates higher ad revenues).

Most people who work in television do not earn these high salaries. The competition for the jobs that carry high salaries is extremely fierce, and the odds of staying on top for long are not very good. Still, there are ways of building toward the top (these are usually based on a combination of persistence, talent, and luck), and, if not everyone makes it to New York or Los Angeles or Chicago or Washington, D.C., there is a great deal to be said for the quality of work (and the quality of life) in Boston, Philadelphia, San Francisco, Pittsburgh, Atlanta, or any one of a dozen other cities that have a reputation for fine performance in broadcasting and cablecasting.

WINNING AWARDS

There are a few dozen annual awards competitions for television productions, and for individual producers, directors, performers, and other purveyors of the television arts and crafts. The best-known of these is the Emmy Award, presented in a variety of categories by the National Academy of Television Arts and Sciences and by its local chapters ("the local Emmy Awards," a competition held annually in most major markets). There are awards for cable programming (the ACE awards), for home video programming (the ViRA), for commercials (the Clio), for news programming (presented by the Columbia School of Journalism), for almost every type of program and every type of person involved with the program.

Winning an award, particularly an Emmy or a Clio, carries a considerable amount of prestige for the winner. A program that wins an Emmy may increase in value, particularly in the off-network syndication marketplace. But most productions, and most producers and directors, never win any awards. This is usually because they do not enter, or because the limited number of categories does not permit a fair competition (for many years, daytime game shows were in the same category as "The Tonight Show," for example), because the essence of their productions cannot be captured in the spot-check scanning that is common in many closed-door judging situations, or, sometimes, because of a built-in prejudice against the productions made by a particular company or station (voting in favor of friends is extremely common, for example). Some producers have a great many awards (because they enter every competition, or be-

cause their productions look good when spot-checked, or, simply, because they are consistently better than the other nominees); others have none.

Do awards have anything to do with real life? Many award-winning producers are out of work. And many producers who have never won an award in their life are earning a very comfortable living. Programs that have never won an award can top the ratings charts, and programs that sweep the awards may be off the air before the competition is complete. So the answer is "no," awards don't have much to do with real life. But it is nice to win recognition, and, if kept in the proper perspective, awards can be a pleasant reward for a job done right. (A show renewal for another season is, however, a far more coveted award than any other.)

NOTHING LASTS FOREVER

There comes a time in every production, regardless of how successful, when the project is complete, when the show goes off the air, when "it's over." Shows go off the air with bad ratings, and, sometimes, with good ratings. Nonbroadcast series cease production because sales projections are not being met, or because a new company president has a better idea. Production companies start up because there's plenty of work around, and close down when competition or marketplace conditions change the rules. Some productions end because of an argument between a producer and a program executive, others because a programmer took a calculated risk and shifted the time slot one half-hour later. Sometimes the audience tires of the show, sometimes the producer or the performer loses interest, and sometimes, albeit rarely, a show goes off at its peak because the key creative people wanted to end production before the series lost any of its magic ("The Dick Van Dyke Show" is the classic example of this phenomenon; "M*A*S*H" did much the same thing years later).

The corollary to "Nothing lasts forever" is "Nobody is indispensible (except, perhaps, the star)." There are many stories of producers, directors, and writers who thought themselves so vital to the success of a show that they demanded either a huge salary or some tremendous perk—and were shown the door. Television is a very competitive business. There are *always* replacements available (perhaps not as good as the originals, but, with a little training, acceptable replacements just the same). The only possible exception here is the star, as NBC has consistently found when Johnny Carson renegotiates contract after contract (Carson, however, is "The Tonight Show"'s third host—the fourth, if you count an early version of the show). CBS was very concerned about replacing Walter Cronkite years ago, but Dan Rather seems to be holding up in the ratings. ABC depends very heavily upon David Hartman for the success of "Good Morning America," where NBC has seen many different performers at the helm of "GMA"'s competitor, "The Today Show." In these situations, and in local news, it's hard to know just how much of a production's success is due to the performer, or the producer, and how much is due to the format. Still, change is the rule, not the exception, and it is not impossible to imagine Carson, Hartman, or any number of "institutions" changing affiliations, particularly in the environment of strategic and business changes that has been so prevalent in the television industry of the 1980s.

WORKING AS AN INDEPENDENT

Despite the visibility of regularly scheduled programming like the nightly news and morning talk shows, most television production is handled on a per-project basis. Even prime time series, most of which appear year round, are produced by a group of independents who work hard for half the year or more, complete the season's requirement, and then move on to other projects until production starts up again for the following year. Many television productions are one-shots, like specials, home video programs, and industrials.

Although the supervisory personnel keep working all year, the project orientation of television production has created opportunities for a great many independent producers, directors, and production companies, as well as an army of free-lancers.

The independent existence requires a special kind of patience, an ability to balance a number of projects at one time, and a talent for developing and maintaining a client base that is sufficiently large to generate a respectable income, and yet sufficiently small to be professionally serviced. Even the most successful independents have a difficult time with the natural ebb and flow of workload; there are times when the office is very, very busy, and there are times when the phone does not ring at all. This situation is as common as it is tricky, particularly when a staff is involved (when times are busy, trained staff people are a real asset, but when things are slow, the overhead can be devastating).

Individual free-lancers have many of the same problems: the work seems to bunch up in certain weeks, and to be nonexistent in others. What's the difference between independent producers and free-lancers? An independent producer usually becomes involved in developing his or her own projects, or has a production office. The free-lancer usually makes a living by hiring out by the day, the week, the month, or by the job (on a flat rate). Some free-lancers are very happy moving from project to project ("It keeps life interesting, always meeting new people and running into new ideas"), while others, usually former staff members accustomed to a regular routine and a regular paycheck, are not all pleased to be gypsies ("It gets me crazy, never knowing who I'm going to be working with, having to negotiate fees every few days, chasing down paychecks, always running into new ways of doing things"). The proper temperament and the proper attitude are essential to the success of a free-lancer. Those who cannot cope either return to staff jobs or leave television altogether.

Most producers and directors have the opportunity to free-lance occasionally, whether they are between staff jobs or simply want a break from the routine. At some point in their careers, most directors free-lance to supplement income or to learn new skills.

CROSSROADS

There are several crossroads that are common to most careers in television production. The first, and easily the most frustrating, occurs during the search for the entry-level job. As explained previously, producers are wont to hire experienced personnel because of the time involved in training, and because they generally have their own established groups. This leads most graduating seniors, and career changers, into an infuriating spin ("He won't hire me without experience, but how can I get experience if no one will hire me?").

There are several ways around this problem. First, work for free. Many colleges and universities offer internship programs with area broadcasters and production companies; many successful producers and directors met their initial contacts through internships. Some producers, or production staffs, can use an extra hand, and unless they're barraged with such offers (which happens in major markets), suffer from lack of space, or problems with liability insurance (employees are covered, and interns are usually protected by a school pol-

icy, but volunteers are a risk). Most of the time, arrangements can be made if the novice is persistent, and if help is needed. Many performers, and producers and directors, got started by "just hanging around the studio," making friends with the production staff, doing odd jobs, sometimes even being part of the audience, arriving earlier, leaving later. High school and college students sometimes find their way into professional situations by emulating them at school, creating their own versions of game shows, news shows, etc. One producer began by building cardboard models of the sets used on all of the NBC game shows; another re-created "The Tonight Show" in his basement every night of the week, using friends as guests. These people really had television in their blood from a very early age.

More ways to get started: if a producer is developing a new project, preparing it for presentation, offer to type, to do research, to work for the experience alone. Try to get the opportunity to meet the producer by writing an article about him or about her or about the show for a magazine or newspaper; most productions are happy to get the publicity. Some game show writers began their career as game show contestants; some people who have appeared on local public affairs programs as guests later find themselves producing the shows; in both of these cases, a combination of personality and persistence helped to separate these people from the many others seen every day by producers and directors.

If these ideas fail, try working in an allied industry. Public relations, advertising, corporate communications, the print media offer parallel experience, and, in some cases, and in some markets, the competition for these jobs may be less severe than it is in television. The third possibility is to try to do something on your own; most people fail at this, but there are success stories of twenty-two-year-olds, and others, who arrange financing for a production that manages to gain some notoriety.

Television is a business that demands resourceful behavior. There are many, many stories of people breaking into the business. It's tough, but it's possible.

The second crossroad is usually found after some time at the "assistant" level: the assistant director, the associate producer, the production assistant who feels ready to make the move but doesn't see an opportunity. An AD tied to a successful director, perhaps a mentor, eventually realizes that the director has no intention of leaving a high-paying, satisfying job; a production assistant who knows that he or she will always be perceived as a PA at the current place of employment; a producer who is second-in-command, but won't be able to take charge unless he or she either plays some heavy political games, or hatches a murder plot against the incumbent senior producer.

In most of these situations, the only solution is to leave. The tricky part usually has to do with the title: although she is doing the job of producer, she still has the "associate producer" title. Making a lateral move in television is usually hard enough, given the competition in most markets, but making a jump requires very special handling. Creating viable options is also a challenge; there are bound to be tempting offers in "parallel areas" (e.g., running the traffic department, becoming an editor, moving to a management position in the operations department), and many of these will carry salaries higher than the cherished producer or director job that is being sought. This is a hard decision; few jobs are more satisfying than producer or director, but most are far more stable, and far less crazy. The choice is frequently a complicated one, for at this point in one's career, the obligations of adulthood may encourage the "safer" choice, while one's own creative needs demand the satisfaction that the producing or directing job can bring.

The third common crossroad is reached after producing or directing, usually on the same show, for many years. The job becomes easy. It is no longer a challenge. It's no longer fun.

Directors at this stage usually overcome the feeling by taking on a few invigorating freelance assignments, changing the routine, or, in some cases, by beginning a parallel career in producing. Producers at this crossroad sometimes try to make the jump to executive producer, or to management, or, in some cases, to directing or writing as a parallel career.

The feeling of complacency is rarely eliminated by changing shows or changing stations. The problem is frequently rooted more deeply; the magic is starting to wear away. Some people find this stage comes at a time when family or friends or outside interests become more important; television people spend too much time working anyway. They change careers at this point, some for good, others for a while, until a new challenge in television comes along. For fast-trackers who reach this stage in their thirties, the solution is oftentimes a new career, or an entirely new plan or motivation. For more seasoned veterans, who believe that forty-five or fifty is "too late to start a whole new career," time either proves them wrong or helps them to find a television situation that is satisfying. Nobody has the answers to these problems; one must ultimately design his or her own career and deal with inevitable peaks and valleys that come along with it.

MAINTAINING FOCUS

Although it is impossible to predict a career path years into the future, it is important to select the proper jobs, to try to build toward an ultimate goal. There are a remarkable number of people, for example, working in New York whose goal is prime time dramatic television, knowing full well that all but a few prime time shows are produced in California.

There are people who have worked their way up to director of a local news program, hoping to direct movies someday, only to find that the proper career route begins with an assistant's job with a film camera operator, or even a secretarial job at one of the studios. Plotting the right steps is difficult but not impossible. And opportunities have a way of finding people who maintain focus.

If you want to produce, get a job with a producer. If you want to work in news, and you can't get a job on a news program, get a job in radio news or at a newspaper. If you want to work on a game show, get to know the names of every game show producer in LA, and let them know who you are and what you can do. If you want to direct television sports, start by covering games in high school or college. Find a way into the booth at a few minor league games (offer to do stats, if they don't have anybody already doing them, or try for an internship). Get involved, any way you can, with people who do what you want to do. And be aware that there are plenty of other people who are doing the same thing you are. The ones who are persistent are the ones who are going to make it.

PERSISTENCE, OPPORTUNITY, AND LUCK

There are many stories, some quite amazing, of how producers and directors have reached success. Some involve talent and persistence; others an abysmal lack of talent and a few lucky breaks. And some just happen without anyone really knowing why or how.

Still, there are steps that every producer, every director, every creative person in television should observe. If they do not guarantee success, they certainly improve the odds.

First and foremost, it is important to keep a close watch on the marketplace. Television is changing, because of cable TV, because of home video, because the networks are changing their strategies, because of government regulations, because equipment is becoming less expensive, because of entrepreneurs and corporate mergers, because of a hundred different reasons. Read the trades, keep your contacts informed of your activities, listen to the grapevine (rumors in the television industry are usually to be trusted), and keep a close watch on both the behavior of the audience (through ratings, etc.), and the behavior of the program executives. In short, find out where the action is, and then adapt yourself and your abilities to fit in.

Second, maintain visibility. Let people know who you are by entering awards competitions, serving on panels and committees, getting your name into the trade press (and, if possible, into the consumer press). Make certain that you get credit for the work that you do (some executives have a reputation for taking the credit "for the staff"), and that key figures in your industry are aware of the quality of your work. Every time you do something new, add it to your reel, and do a mailing, if it's appropriate in your corner of the business. Some producers and directors are particularly skilled in gaining recognition; observe them carefully, copy what you need, and then develop your own style. But don't keep your work a secret, or assume that interested people will see it on the air. They probably won't.

Third, and this sounds pretty obvious, do good work. Although quality production alone is not the key to success, it is the most important element in the formula, particularly for those who possess above-average talents. The quality of the finished production, whether it's live or on tape, a commercial, an industrial or a network show, is judged by the viewer in terms of entertainment or information value. Within the industry, however, work is judged by additional criteria: adherence to schedule and budget, treatment of the people involved, and the general atmosphere created. It is these factors that insure long-term success, and careful attention paid to them should garner the one compliment that tells a producer, or a director, that he or she is doing the job right.

When the guy sweeping the floors looks up after a show, says it was a good show, and compliments you on "being a real professional," you know you've done the job right. And for all the awards, and all the raises and bonuses, that's what it's all about.

GLOSSARY

Many of terms used in television production vary from city to city, from facility to facility. The list that follows contains most of the common terms; it should be amended and updated periodically by the reader.

Above-the-line   All budget items paid to creative personnel, including the producer, director, writer, production staff, and talent. Rights payments and agency commissions are also considered to be above-the-line items.

AD   Associate director or assistant director. The director's right hand, the one who takes care of details so that the director may concentrate on the overall project.

ADO   Proprietary name for a digital video-effects device manufactured by Ampex, which permits the user to create three-dimensional images on the television screen.

AFTRA   American Federation of Television and Radio Artists. One of two performers' unions (the other one is SAG, the Screen Actors Guild).

Agent   Someone who represents the interests of a producer, director, writer, performer, or other creative artist, who promotes interest in their services, receives proposals and advises the client to select the best ones. An agent may also negotiate on the client's behalf. Agents may be individuals or staffers at an agency (like the William Morris Agency or ICM).

Ambience (also called **ambient noise**)   Atmospheric sound heard at any location. Examples would be street noise, the commotion of a crowd at a sporting event, the strained silence during a courtroom trial. Also used to describe the overall feeling of a stage setting.

Ancillary markets   Areas of profit potential outside the normal range of broadcast and pay television markets, usually including home video, schools, military bases, and libraries.

Ancillary rights   Rights related to the ancillary marketing of a television property, and the right to create products (like toys, games, apparel, stuffed animals, and greeting cards) based on the property.

AP   Associate producer. The producer's right hand, usually responsible for details.

ASCAP   American Society of Composers, Authors and Publishers. One of two clearance houses for music publishing rights (the other is BMI).

Aspect ratio   The ratio of height to width, as on a television screen. A 3:4 ratio is standard in television; a ratio closer to 3:5 is common in the movies (which is why, for example, title sequences on many motion pictures appear to be compressed when seen on television).

Assembly (also called **assembly editing**)   The process of editing pieces in a consecutive order, where A is followed by B which is followed by C (see also **insert editing**).

Audio   The sound portion of a television program, including spoken words, sound effects, music, and ambient sound.

Autoassembly   The virtually automatic on-line editing of a program or commercial, wherein time code numbers are loaded into the editing computer (via keyboard input, or retrieval from floppy disk) and, in theory, the machinery edits the production automatically.

Back end   Profits generated by a production in ancillary markets. The term is commonly used in the phrase "back-end deal," where the beneficiary recieves profits only if the production makes money in the ancillary markets.

Back-timing   Counting the number of minutes or seconds remaining in a live or live-on-tape program. With this information, the producer or director can revise the format so that the program ends on time.

Barn door   One of four metal flaps mounted on the front of a studio light to define its beam.

Bars   See **color bars.**

Barter   A form of broadcast syndication. Local stations receive the program at no charge, provided that a certain number of commercials supplied with the program are run intact (the station may sell all other commercial spots and keep the money).

Barter + cash   As above, but the syndicator (usually on behalf of an advertiser) not only supplies the program for free, but pays the station a fee as well.

Basic cable   The group of cable channels offered to subscribers at the lowest monthly fee. "Basic cable services" include all broadcast television stations in the area, plus some or all of these cable network services: WTBS/The Superstation, Cable News Network, MTV/Music Television, The USA Network, The Weather Channel, CBN, and a variety of text services (classified ads, weather reports, channel listings, etc.).

BBC   British Broadcasting Company. Government-subsidized network television in Great Britain, widely known for superb programming. Some BBC programs are seen on PBS, frequently with wraparounds made by American broadcasters. Examples include "Masterpiece Theater" and "Mystery."

Beauty shot   A wide-angle shot of an entire studio, studio set, or location; also, a "pretty" presentation of a product, usually at the end of a commercial.

Below-the-line   Budget items required for the manufacture of a television production, including the facilities, the crew, the sets, the costumes, the travel expenses. Realistically, any items not con-

sidered to be **above-the-line** are considered to be **below-the-line.** The definition varies with each situation.

Beta   Sony's proprietary name for their ½-inch videocassette tape format, primarily used in the home (as Betamax), gaining popularity as an ENG format (as Betacam) as well (see also **VHS**).

Billboards   Full-frame shots of performers, products, or sponsors, typically used at the start, or the end, of a major event (or any other show).

Black   The absence of a television picture; a dark screen; the black areas in a television picture.

Blocking   Placing performers, and cameras, in their proper positions for each scene.

BMI   Broadcast Music, Inc. One of two clearance organizations for music publishers in America (the other is ASCAP).

Booker   A staff member who arranges for guest appearances, usually on a talk show.

Booking board   A wall-sized calendar, where guest names are inserted, show by show, segment by segment, by the booker.

Boom   A long pole, used to suspend a microphone above a performer, just out of camera range. A "fishpole boom" is usually hand-held; larger booms are supported by upright beams, some quite complex in their capacity to move the microphone itself.

Broadcast   The transmission of television (and radio) signals through the air.

Bug   The union seal that is stamped on the back of sets made in a union shop, or, in some cases, on the label of every tape made in a union shop.

Build day   The day scheduled for putting up the set in the studio; also called a set day or a setup day.

Burn (or **burn-in**)   A permanent or long-lasting mark on the image from a camera that has been trained on an extremely bright light or a very bright area contrasting with a dark background. When the camera tube on an older camera is burned, it may have to be replaced. Newer tubes are somewhat more resilient.

Bus   A row of buttons on a television switcher, used to select the program material ("the program bus"), the input to a preview monitor ("the preview bus"), or to set up special effects ("the special-effects bus").

Buy take   A take that is approved by the director for use in the finished production (the word "buy" is probably short for the director saying, "I'll buy that").

Cablecast   The transmission of television (and radio) signals through a cable that is wired directly into the subscriber's television set.

Call   The time that performers, technical crew, or production staff are required to be on the set or on location, ready to start working.

Camera left   Left as seen from the camera's point of view, also known as **stage right**.

Camera right   Right as seen from the camera's point of view, also known as **stage left**.

Cart   An audio cartridge, commonly used for the recording and playback of music, sound effects, and voice-overs. With the touch of a button, a cart will automatically find its starting point ("the cue spot") through the use of a tone above the range of most people's hearing.

CCU   Camera control unit. Attached via cable to the camera head, the CCU allows the operator (usually a video engineer) to adjust color, contrast, and related technical points before or during the production without actually disturbing the camera itself.

Character generator   A computer whose keyboard is used to type letters, numbers, and design elements onto a television screen.

Charts   (1) In music, the arrangements on paper. (2) "Camera charts" (the word "camera" is often omitted) are used to set registration and other technical points prior to the start of shooting.

Chroma   Or, more accurately, the "chromanance" portion of a video signal. Video operates with three chroma channels, red, green, and blue. The combination of each of these channels, in varying degrees, creates every color seen on television.

By increasing the red chroma level, the entire picture will appear more red than before.

Chroma-key   The replacement of one color (usually a saturated blue or green) from a television picture in favor of input from a second television picture.

Chyron   Proprietary name for a popular character generator system.

Clearance   (1) Permission to use music, photos, or footage in a television production, granted by the individual or company who owns the copyright; also called a "rights clearance." (2) In syndication, a clearance is a sale to a station in a particular broadcast market ("We've cleared Kansas City").

CMX   Proprietary name for a popular computerized editing system.

Color bars   An engineering test pattern for color, also called "bars." When color bars are recorded onto the head of a tape for adjustment during playback, a test tone is frequently recorded as well, hence the term "bars and tone."

Contingency   When preparing a production budget, it is best to add about 10 percent for emergencies. This emergency fund is called the contingency.

Continuity   (1) The scripted or produced material used to connect program elements; (2) the flow from one cut or one segment to another. When one is "checking for continuity," he or she is likely to observe what is being said, the positions of the performers, the positions of certain props and set pieces from one piece to the next. Everything must follow, everything must look right. Any problems in the program's flow are called "continuity problems."

Control track   The videotape track used exclusively for a signal that synchronizes the picture and the sound tracks.

Cookie   A cutout placed in front of a spotlight to project a pattern onto a cyc or set piece.

CPB   Corporation for Public Broadcasting. A major source of financing for PBS, consolidating many contributions from government, foundations, and private business.

Crane   A large camera support that permits movement to high (over ten feet) and low (under one foot) positions. It looks and operates like an outstretched arm, palm up, with the camera mounted on the palm.

Crawl   Sometimes used as a synonym for **credits** (an older use of the term). On most character generators, the crawl function causes letters to move from one side of the screen to the other (as in a news bulletin at the bottom of the screen).

Credits   The on-screen listing of people associated with a production. Performers are usually listed in the opening credits (so are the writer, producer, and director on most fiction projects); everyone else is credited at the end of the program. Some productions do not include credits.

Crew   The technical, engineering, and staging personnel.

Cue   A signal to begin. A cue to a performer is a signal to start moving, or to start talking. A camera cue is a signal from director to TD to switch to a new camera. When playing a tape, the tape is "cued" when it is at the starting position.

Cue light   See **tally light.**

Cut   (1) An instantaneous change from one image to another, or (2) an instruction from the director to stop the action in the studio.

Cyc   (Short for cyclorama, a term that nobody uses much anymore.) The infinite background found in most studios that stretches from floor to grid. It may be a solid wall that curves into the floor to create a seamless sweep (a "hard cyc") or a stretched fabric with a scooplike transition from floor to wall (a "soft cyc").

Dark   When a studio is not being used, it is said to be "dark." The term has extended to programs that are normally produced on a regular daily schedule; on days when the program will not be produced, the schedule reads, simply, "dark."

Dead roll   The process of starting a videotape (or film) program at its appointed time, and then waiting for a live event to end (e.g., a football game that's running late) before joining the program in progress.

DGA   The Directors Guild of America, the union that represents directors, associate directors, and stage managers in many markets.

Dimmer   A lighting control that permits varying degrees of brightness.

Disc   Shorthand for slow-motion disc.

Disclaimer   An audio or visual announcement that limits, for example, entry in a contest for employees of the company, its advertising agency, etc., or the producer's liability related to the showing of a particular sequence (e.g., a cure for the common cold, "which may not work for everyone").

Dissolve   A gradual transition from one image to another.

Dolby   Proprietary name for a noise reduction system, usually used to eliminate hiss from tape recordings.

Dolly   (1) A camera platform with particularly accurate control over movement as a result of a special wheel and drive mechanism; (2) the director's instruction to move the camera, on the dolly, to the left ("Dolly left"), right ("Dolly right"), forward ("Dolly forward" or "Dolly upstage"), or back ("Dolly back" or "Dolly downstage").

Double system   The recording of video on a videotape recorder and audio on a separate audiotape recorder. The machines are usually tied together with a synchronizing unit.

Downlink   A center for the reception of incoming signals from broadcast satellites, which includes a dishlike antenna as well as the necessary support equipment. Also called an earth station.

Downstage   Toward the front of the stage.

Downstream   Images or effects added after the principal areas of the switcher have been used; the completed images are headed "downstream," toward the VTR, master control, or the transmitter.

Downtime   Time when equipment is inoperable ("The equipment is down"), usually deducted from the hourly facilities rate.

Dropout   A loss of picture, usually caused by imperfections on a videotape (which may be caused by overuse). Dropout is usually seen as tiny white/silver flecks on the video screen.

Dry run   A basic rehearsal, usually without some elements (e.g., cameras, props, costumes, lights, any number of things).

Dub or **Dupe**   A copy of a video or audio tape.

DVE   Digital video effects. Generic term for computerized special-effects devices.

Earth station   See **downlink, uplink.**

ECU   Extreme close-up.

Edit decision   The individual elements on an **edit plan**.

Edit plan   A specific list of "in" and "out" points, usually keyed to time codes and reel numbers, prepared by a field producer. This list is used by the editor to rough cut without supervision.

EFP   Electronic field production. Location video.

Elephant doors   Large studio doors ("large enough to fit an elephant through them").

ENG   Electronic news gathering. Location video, usually with very lightweight equipment and some capacity for live transmission.

Equalization (or **EQ**)   Modification of certain audio frequencies to create a more pleasant, more realistic rendition on the sound track.

ESS   Electronic still-frame storage device.

ESU   Electronic setup. Time at the start of each production schedule allotted for the crew to set up cameras, microphones, and other equipment, and to run the appropriate technical tests prior to FAX.

Fade   Dissolve either up from black ("fade in") or down to black ("fade out").

Fader   (1) On an audio mixer, a volume control for any one of the inputs. (2) On a video switcher, the bar used to move from one bus to another, as in a dissolve or a wipe.

FAX   (1) The form used in many local stations to order facilities. (2) The time on the production schedule when the entire studio facility has been properly set up and tested, and is now available to the director.

Film chain   One or more film (or slide) projectors, connected via mirrors and electronic synchronization equipment to a television camera. Also called a **telecine**.

Final cut   The right to declare a project complete; that is, the right to make the final editorial decisions.

Fine cut   A relatively polished version of an edited production.

Fishpole   A small boom, for a microphone, suspended above the action by hand.

Flag   A piece of cardboard, metal, or cloth stretched on a metal frame used to shield certain areas from a beam of light. Usually attached to a light stand or taped to a set piece.

Flat   A rectangular piece of scenery, usually four feet wide and eight or twelve feet high, made of either wood or canvas that is stretched on a wooden frame. A flat is usually painted on one side and supported by a "jack" (which makes the angle from the floor to the flat needed for support) with a sandbag for ballast.

Floodlight   A light source that illuminates a broad area.

Floor   Shorthand for the floor of the studio ("The director is working on the floor").

Floor manager   See **stage manager.**

Foldback   Studio public address, usually as an audio monitor for use of performer or audience.

Follow spot   A large spotlight, which bathes the performer in a pool of light (reminiscent of vaudeville theaters).

Format   (1) The show's routine, which lists all of its component segments. (2) The type of tape used for recording or playback (e.g., Beta format, 1-inch format, U-Matic format).

Frame   A single television image (there are thirty frames per second on television). Half a frame, showing only half the scanning cycle, is called a field.

Frame store   A digital recording of a single frame of video.

Free-lance   Employment on a per-project basis. In order to maintain the freedom to choose projects, many producers, directors, and writers offer their services on a free-lance basis.

Freeze frame   A single frame of video, shown as a still picture.

Full track   Audiotape recorders are usually set up to record four tracks on a single tape (your home audiocassette recorder uses a pair of stereo tracks in each direction for a total of four tracks). When a recorder uses all four of these tracks as one single track (thus using a larger surface area and improving the quality of the recording), a full track recording is made.

FX   Abbreviation for "effects." ("SFX" can mean either "sound effects" or "special effects.")

Gaffer   A stagehand, specializing in lighting.

Gel   A piece of colored plastic attached to the front of a lighting instrument to color its beam.

Generation   When a dub is made from a master tape, it is called a first-generation dub. When a copy is made from the first-generation dub, it is called a second-generation dub. The quality decreases with each generation of recording.

Gobo   Any graphic or set piece designed for shooting through, as if it were a frame. Frequently used to establish three-dimensional space.

Gofer   As in "go fer coffee" or "go fer a copy of today's paper." An entry-level PA, who takes care of errands and other such chores.

Green room   A waiting area for studio guests.

Grid   The crossed network of pipes and electrical outlets suspended from the ceiling of a studio. Lighting instruments (and, occasionally, microphones) are hung from the grid.

Grip   A stagehand.

Half track   A recording where half of the tape's surface is used (see also **full track**).

Hard cyc   See **cyc**.

Head end   The master control room at a cable television operation, where incoming signals are assigned channel numbers and then transmitted to subscribers.

Headline budget   A budget summary that shows the totals of each of the major areas (staff, talent, facilities, etc.), but does not show the detail.

Headroom   Unused space at the top of the television screen. When an object appears to be too close to the top of the screen, a director will ask the camera operator to recompose with less headroom.

HMI   Halogen-metal-iodide. A lamp that provides illumination closely resembling daylight.

IATSE   International Alliance of Theatrical Stage Employees. One of several unions that represent technicians, engineers, and stagehands.

IBEW   International Brotherhood of Electrical Workers. A union for technicians and engineers.

ID   Station identification, sponsor identification on camera, etc.

IFB   Internal feedback. A small earphone, frequently used by newscasters on live events, usually wired directly to the producer or the control room for fast-breaking information.

Insert editing   The insertion of piece C between A and B, or, more practically speaking, the insertion of a shot or scene within the body of another scene (see also **assembly**).

Interactive video   Television that allows manipulation of its playback sequence by the viewer, usually with special hardware (e.g., a videodisc player or specially equipped computer device).

Interstitial   Material seen between programs. The term is commonly used in the pay TV services, where movies may not end on the hour or half-hour, and "interstitial programming" is used to fill the gap until the next program.

ISO   Short for "isolated camera" or "isolated feed." By placing cameras on ISO, the director does not make final cutting decisions during the event, but in the peace and relative calm of the editing room instead.

JIP   Abbreviation used at local stations to mean "join-in-progress."

Jump cut   When two pictures are visually similar, they should not be cut to follow one another. A "jump cut" causes the action to appear to jump from one frame to another, which is disturbing to the eye (and generally considered to be a slipshod technique).

Key   The elimination of part of one image to replace it with place lettering, a title, or another picture element from another image.

Key light   The principal illumination on a performer, set piece, prop, etc.

Keystone   A graphic that should appear rectangular on camera, but appears to be angled back right or back left, like the shape of a keystone.

Laserdisc   An interactive videodisc system that uses a laser beam as a stylus.

Lead-in   (1) The program that precedes another (e.g., "Local news provides an excellent lead-in for 'The CBS Evening News'"); (2) a scripted introduction to a guest, a segment, or a videotaped report.

Lead-out   The scripted conclusion that follows a segment.

Lens line (prompter)   A teleprompting system that shows the line currently being read by the performer on a line that appears at the center of the lens.

Letter agreement   A letter of intent, frequently used as a short form for a contract.

License   The right to use a copyrighted product, for a specific purpose, within specific time guidelines. A television program "licenses" a program for use on the air in exchange for a "license fee." A producer would "license" a piece of music or a story, usually in exchange for a fee paid by the production.

Lighting plot   A map drawn by the lighting director that shows the position of every light on the grid in relation to the set, the studio floor, and, in some cases, the paths of performers' movement.

Limbo   A set piece that is placed in front of the cyc, and has no definable background.

Line monitor   The control room monitor that shows exactly what is being fed live or to the tape machines. Also called the program monitor.

Lip-sync   Moving one's mouth in careful synchronization with, for example, the lyrics of a song.

Lockup   The time required for a videotape machine to start showing clean pictures; some machines require several seconds of play before they lock up, others lock up immediately.

Log   A computer printout of every program, commercial, promo, public service announcement (PSA), and station ID and the exact times they are to appear on the air.

Logo   The official, or unofficial, trademark for a program, station, network, or a product, usually seen in advertisements, and at the opening and closing of the program. The CBS "eye" is a well-known example of a logo.

Luminance   The relative brightness of a television image, made possible by the combination of a black-and-white channel with the chromanance channels (which provide the color). It is the luminance channel that allows color television pictures to be seen on black-and-white TV sets.

Magazine format   The television version of a magazine, where several videotaped stories are linked together by one or two hosts and some snappy graphics ("P.M. Magazine" is a good example).

Major market   The ten or fifteen largest metropolitan areas in the U.S. are considered to be the major markets. These include New York City, Los Angeles, Chicago, Philadelphia, Detroit, San Francisco–Oakland, Washington, D.C. (including Baltimore), Dallas–Fort Worth, Houston, Boston, as well as cities like St. Louis, Pittsburgh, Minneapolis–St. Paul, and Atlanta.

Manager   An artist's representative, usually one who concentrates on the business side (as opposed to the promotion side, which is typically the role of an agent).

Market   In local television, the area that receives the signal from a television station with an ordinary antenna. New York is the no. 1 market because signals from New York stations reach more homes than in any other market.

Master (tape)   The original videotape, whether it is made by a camera (the "camera master") or in an editing room (the "edited master").

Master control   In a local station or a network, the control room that produces the on-air signal or feed.

Match cut   A cut from one image to another, when the images are carefully lined up to match a portion of one picture to another (e.g., a spinning record to a rotating wheel).

Matte   One of several systems that allows the cutting of portions of one image out, and the replacement of portions of another image in their place (see also **key**).

Meal penalty   When a member of a union is denied a meal break at the scheduled time (with a certain amount of flexibility, of course), or breaks later than scheduled, the production is expected to pay a penalty fee.

Minicam   Another name for an ENG camera.

Miniwave   The microwave connection of a camera to a control room without the use of a cable; commonly used on multicamera remotes where cable runs are difficult.

Mirage   Proprietary name for a digital video-effects device (see also **Quantel, ADO**).

Mixdown   The playback of multiple audio tracks at varying levels, carefully monitored and modified by an engineer, into a single recording track.

Mixer   (1) An audio control board that permits the balance of two or more inputs (e.g., microphones, videotape playbacks) and the output of a single mono or stereo signal. (2) One who operates such a device.

Mobile unit   A truck or van designed to be a control room on wheels, generally supplied with several cameras and, on special request, microwave transmission equipment.

Monitor   (1) In video, a television screen that shows a selected video signal (e.g., the program feed, the preview feed, the feed from one camera). Monitors are usually of higher quality than standard television sets, and do not normally contain any audio circuitry (nor do they contain any reception equipment, so they cannot be used to watch regular television without connection to a tuner). (2) In audio, a speaker that reproduces a selected source of sound (e.g., the program sound track, a song for lip-sync).

Montage   (1) An edited sequence of pictures that creates an impression (of a place, a person's background), usually played to a musical sound track. (2) A proprietary "picture processing" system that brings some of the ease of word processing to videotape editing.

MOS   Without sound (or, supposedly from the old German directors in Hollywood, "mit out sound").

Multitrack   Multiple audio tracks, as recorded on an audiotape recorder with eight, sixteen, twenty-four, or thirty-six tracks, and as mixed on an audio console with a comparable number of individual tracks.

NAB   National Association of Broadcasters. The NAB sets codes and standards that are generally followed by broadcasters nationwide. It also sponsors an annual convention where the latest in television equipment is shown and demonstrated. Also a lobby group.

Narrowcast   Transmission to a limited area (e.g., cable subscribers who have paid a fee to see a program).

NCTA   National Cable Television Association. Cable television's version of the NAB.

NTSC   Technical abbreviation for the color television system used in the United States, Canada, and in several other countries. NTSC employs 525 scanning lines to create a television picture, whereas PAL (the system used in Europe and many other countries) uses 625, which is one reason why American television sets cannot be used abroad, and why the picture on a European set looks so much clearer than the picture on a comparable American set.

Off-line (editing)   The video equivalent of rough cutting in motion pictures. Editing on ¾-inch videocassettes is commonly called off-line editing, even if the finished product will be mastered on ¾-inch tape. (A new term, "off-off-line," is creeping into the lingo as a term for rough-cutting on ½-inch VHS or Beta tape.)

Off-network   Programs originally seen on ABC, CBS, or NBC, now available for play on local stations or on cable (also called reruns).

Omni-directional   Microphone pickup pattern with equal sensitivity in all directions.

On-line   Editing on 1-inch or 2-inch videotape, generally with the use of a computerized system, to create a finished master, ready for air.

Outcue   The last few words, sounds, or pictures on a videotape, usually the last five seconds, used by the director to ready the control room for the next program element.

PA   (1) Production assistant; (2) public address system, used in the studio or on location.

Pacing   The intangible feeling that an appropriate amount of time is allotted to each program element, and that the camera cuts and music are in the right places. Similar to "a sense of timing," but usually applicable to entire shows.

Pad   Extra time at the end of the show, during which the performer must "pad" or ad-lib until time is up.

Paint box   A digital graphic arts device, which permits the artist to work directly on the television image.

PAL   The European television system (see **NTSC**).

Pan   Move the camera from right to left ("pan left"), or from left to right ("pan right").

Paper cut   An edit plan that includes every in and out point, keyed either to time codes, running times, or cues, used by the producer to make most of the editing decisions before entering the editing room.

Participation   See **profit participation**.

Pay-per-view (also called **PPV**)   The transmission of a specific program, usually a major event, to an audience who has paid a fee to see it live.

Pay TV   Cable or broadcast channels available only to those who pay a subscription fee, usually on a monthly basis.

Pedestal   (1) The heavy-duty camera base used in a studio, which contains a hydraulic lift that makes raising and lowering the camera easy. (2) The basic black level of a television picture, as measured on a waveform monitor.

Pension and welfare   Contributions required by union contracts as benefits for their members, usually paid as a percentage surcharge above the hourly or weekly fees.

Performance area   A specific area on a talk or variety show set where performers sing. On "The Tonight Show Starring Johnny Carson," the performance area is located in between his desk area ("the interview area") and the band, and is covered by a drape during his opening monologue.

Pickup pattern   The shape of the area covered by a microphone.

Pixel   The smallest definable component of a television picture. The term is frequently used in connection with digital-effects devices, and with a paint box, to describe the tiny elements that make up any television image.

PL   Private line. A studio intercom system, usually employing headsets.

Playback   The playing of videotape or audiotape.

Plot   (1) The story line of a fiction program. (2) The lighting director's plan for the setup of lighting instruments.

Plug   A mention of a commercial product or event on the air, a promotional announcement.

Pot   Short for potentiometer, a term nobody uses anymore. A volume control.

Pre-roll   The amount of time required for a videotape machine to stabilize and show clean pictures (usually expressed as "five-second pre-roll").

Preset   The arrangement of an effect on a video switcher prior to its use.

Preview bus   One of several rows of buttons on a video switcher. This one controls the feed to a **preview monitor** (see below), and is also used with other buses to set up special effects.

Preview monitor   A control room monitor used to set up shots and special effects before they are taken for air.

Profit participation   A contractual promise to be paid a percentage of profits generated by a television program (or movie, etc.), usually subject to all sorts of terms and conditions.

Promo   A promotional announcement, whether scripted and read by an announcer or prerecorded on videotape.

Prompter (also, **TelePrompTer**)   The mechanical or electronic equivalent of cue cards. Several systems are currently in use (see chapter 6 for descriptions).

Prop   Short for "properties." Any piece on the set that is handled during a performance. (Small set pieces that are not handled are correctly called set decoration or set pieces, but they're usually called props as well.)

Protection master   A copy of a videotape master, usually locked away in a safe place and used only if the original tape is damaged or lost.

PSA   Public service announcement. A promotional announcement, commercial, or plug given to a community organization, charity, or other nonprofit body as a gesture of goodwill by local television stations.

Pull back or **pull out**   Directing lingo for "zoom out."

Push   Directing lingo for "zoom in" (also "push in").

Quad split   A switcher effect, which permits four different images to appear on the screen at once.

Quantel   Proprietary name for a digital special-effects device (see chapter 7).

Rack focus   Changing point of focus, usually from a near subject to one that's far away (or vice versa).

Rating   A measure of the success of a television program, given in points. The rating is actually the result of dividing the number of households tuned into a program by the total number of television households in a given area (see also **share**).

Read-through   The performers' first reading of script, usually in a group session guided by the director.

Rear-screen projection (RP)   Film or slide projection system that shows the image through a translucent screen. The projector sits behind the screen, where it cannot be seen on camera. The proprietary name "Vizmo" is sometimes used as a synonym. The technique, most often used to show graphics behind a news anchor, has been supplanted by chroma-key, Quantel, and other special-effects systems.

Recoupment   The point at which a production's earnings are equal to the monies invested in its development, production, and marketing. The point of recoupment is important because profit participation generally begins after a project has "recouped" its investment.

Registration   Aligning the three color camera tubes so that they all form one clear image (a picture that is not registered properly will show a slight red, green, or blue outline on portions of the image).

Release   Or "release form." An agreement signed by anyone who appears on camera, granting permission to the producer, station, or production company to show the tape.

Remote   A production outside a studio facility, usually one with multiple cameras.

Remote truck   See **mobile unit**.

Residual   A payment to a performer, director, writer, or, sometimes, producer, when a commercial or other production is played a number of times beyond the number covered by the original fees. Residuals are part of many union agreements, but are negotiated in others.

RF Technically speaking, these initials are short for radio frequency, the portion of the electromagnetic spectrum allotted to broadcasting. In production circles, the term "RF" has become shorthand for a wireless microphone.

Room tone The ambient sound in an empty room. Every location has its own tone, and the differences between rooms can be heard during quiet portions of any audio recording. A good audio engineer will record a minute or two in the quiet of every room where production is done, for later use filling in quiet times during the edit and the sound mix.

Rough cut The first edited version of a program or commercial.

Rough mix An early mix of audio tracks, usually produced to provide a general sense of the sound of the finished sound track.

Routine See **rundown.**

Royalty A negotiated percentage of profits. An up-front fee may be considered to be an "advance against royalties," meaning that the producer or investor must earn a specific amount of money before royalties are paid.

RP See **rear projection.**

Rundown A list of program elements in the order they are to be recorded or shown in the completed production.

Running time The amount of time elapsed since the start of the program or commercial, or the total time of the production as a whole.

Run-through A rehearsal, usually without certain production elements (cameras, costumes, makeup, audience, etc.—it varies with each situation).

SA Studio address system. The public address system that allows the director to speak into a microphone in the control room and be heard throughout the studio.

SAG The Screen Actors Guild. One of two performers' unions in television; the other is AFTRA.

Scale Minimum payments, as established by unions and guilds for their members.

Scoop A common studio floodlight, usually suspended from the grid to provide general illumination on the set or audience area.

Scout A location survey.

Scratch track A temporary audio track, such as narration, laid down by a director to pace the editing of the visuals, later replaced by the real thing.

Script girl (boy) A production assistant or AD who pays close attention to the script, and to continuity, and takes tape notes as well.

SECAM The French television system, used in Africa and other countries. Roughly comparable to PAL.

Second unit The director and crew responsible for producing background footage, extra interview material, or special action sequences involving stunt doubles instead of principal players.

Segment time The amount of time allotted to a specific program element.

Segue A smooth transition from one picture, or one sound, to another.

Set (1) Scenery; (2) the process of placing scenery in a studio.

Set and light The process of placing the scenery in a studio and lighting it (also "build," to put up the set in the studio).

SFX (1) Special effects; (2) sound effects.

Share A measure of success of a television program, given in points. The share is actually the result of dividing the number of households tuned into a program by the total number of homes using television in a given area during a specific time period. Since the program's share (also, "audience share") is based on actual viewership (as opposed to homes who own television sets, the basis of ratings), share points are usually more meaningful than ratings points (see also **rating**).

Shotgun A microphone designed to pick up sound sources located some distance from the microphone position. Shotgun mikes are quite sensitive, but only in a very targeted, very narrow pickup pattern.

Shot sheet A list of camera shots, usually prepared by the AD, which permits the director to call for shots by number (e.g., "Camera two, shot number thirteen" instead of "Camera two, get the right anchor, now pan left and widen to include the monitor").

Slate A blackboard (or erasable plastic board) used at the head of every recorded take to identify the name of the production, the producer, the director, the scene number, the take number, and any other information that may prove useful in editing (e.g., script page numbers to be included in this take).

On studio productions, the slate may be created on a character generator (provided that it does not cause confusion with the use of the character generator during the production itself).

Slo-mo Slow-motion playback, either via 1-inch videotape recorder, slo-mo disc, or digital device.

SMPTE Society of Motion Picture and Television Engineers, the organization that developed time codes (which is also called "SMPTE" or "SMPTE code" in some facilities). The pronunciation is usually "simp-TEE."

SOT Sound on tape. When the tape is played, there is sound recorded on the track. SOF is the film equivalent.

Sound bite A sound recording, usually a short one that will be used as part of an edited piece.

Speed-up An instruction to a performer to read or move faster.

Spot (1) A lighting instrument that throws a highly focused beam (also called a spotlight). (2) A commercial, usually sold either singly or in small lots to a sponsor.

Stagehand A member of the crew responsible for the physical handling of set pieces, props, and lighting instruments. Generally a union position in the larger markets and at the networks, the actual responsibilities of the stagehand vary with each facility and each market.

Stage left From a performer's point of view when he or she is facing the audience, the left side of the stage (which is **camera right**).

Stage manager The director's eyes, ears, and voice on the studio floor, responsible for all of the stagehands, the scenery, the cuing, and communication with performers, etc.

Stage right From a performer's point of view, the right side of the stage (which is **camera left**).

Standards and Practices A network department responsible for maintaining good taste on the air; television's version of the censor.

Star filter An attachment to a camera lens that causes points of light to glow like stars.

Steadicam Proprietary name for a hand-held camera mount, supported by a brace, which permits the operator to hold the camera steady even under adverse circumstances.

Sticks A comparatively portable wooden tripod used on location.

Stock (1) Videotape, usually blank ("raw stock"). (2) Footage, from a library, as in "stock footage."

Storyboard An image-by-image sketch of a production, usually a commercial, which adds pictures to the scripted words. Also called a board.

Stretch The signal to a performer to "kill some time," because the program is running ahead of schedule, or because the next program element is not ready.

Strike Dismantle a stage set, remove a specific set piece or prop.

Super Short for "superimpose," or "superimposition." Placing one image on top of another, so that the images are somewhat transparent.

Switcher The video switching console, and, occasionally, the person who operates it.

Sync Synchronization. Usually used to describe the running of sound and picture in unison.

Sync right (synchronization right) The right to include a musical composition in a television production. This right deals only with the publisher's permission (on behalf of the composer); the artist grants a separate "performance right," which permits the use of the performance.

Take (usually followed by a number)   A reference device used to identify recordings of a scene or a part of a scene (see also **buy take**).

Talent   Generic term for all performers.

Talkback   A public address system, with microphone in the control room, which permits the director to speak to the studio.

Tally light   The little red light on top of a television camera that glows when the camera is on. (It's also called a "cue light," because performers are sometimes told to "start talking when the red light goes on.")

Tape log   A "table of contents" for a videotape, which should be typed and included in every box of recorded videotape for later reference.

TBC   See **time base corrector**.

TD   Technical director, the person who operates the switcher and, in some cases, operates as the senior technical person on the production, or as the crew chief as well.

Tech reqs   Technical requirements, the list of all technical engineering elements needed for a production (pronounced "tek-reks").

Telecine   Strictly speaking, a television camera designed as part of a film or slide playback system. The term "telecine" has become a synonym for the entire system, which permits film footage or slides to be shown on television (pronounced "tel-eh-SEE-nee").

Teleconference   A meeting held by television. Participants in different cities (or countries) sit in specially equipped meeting rooms or studios and speak to one another via microwave or satellite.

TelePrompTer   A proprietary term for a device that shows the script scrolling above, below, or in front of a camera lens. The term has become somewhat generic for all such prompting devices.

Time base corrector   Frequently abbreviated "TBC." An electronic device that insures picture stability during the playback of ¾-inch videotapes. It is not essential for office playback, but it must be used for broadcast and for most editing situations.

Time code   An electronic frame numbering system that assigns every frame of videotape (there are thirty frames per second) a number. Time codes may be recorded as "time of day" (e.g., the actual time that the recording is made) or from zero (e.g., the first time code will read 0 hours, 0 minutes, 0 seconds, 0 frames).

Top   The beginning, as in "Take it from the top," or "There are some problems at the top of the show, but the rest is fine."

Treatment   A script synopsis, usually two to ten pages long.

Trifold   A piece of scenery built of three flats, two of which are connected at the seams. A trifold flat can be used to create a room in seconds (there is no fourth wall in most situations).

Truck left   Director's instruction to camera operator to move the camera and pedestal to the operator's left.

Truck right   As above, to the operator's right.

TV safety   The 80 percent of a television screen that's in the center. The extra 10 percent on the top and bottom and on the sides are said to be "outside TV safety," which means that some of the information in these areas may not be seen on some home television sets.

Twofold   A piece of scenery built of two flats, joined together. This "V-shaped" combination can create the corner of any room, which may be enough set for an interview, for example.

U-Matic   Sony's proprietary term for ¾-inch videotape equipment, sometimes used as a generic term for the format.

Uplink   A dish-type antenna and related electronic gear that sends satellite signals up, toward the satellite. Also called an "earth station."

Upstage   Toward the back of the stage.

VCR   Videocassette recorder.

Vectorscope   A technical instrument that measures and displays information about video color.

VHS   A ½-inch videocassette format (the letters originally stood for "video home system," a term that is rarely used today). See also **Beta.**

Video   The picture portion of a television signal. See also **audio.**

Video art   A relatively new art form that uses video technology as its medium.

Viz-code   Also, "visual time code." Time code that is "burned" into a window on videotape dubs for easy reference to original materials that will be used in editing.

VTR   Videotape recorder.

VU meter   A meter that monitors sound levels in decibels (dBs).

Walk-through   Similar to a read-through or a run-through; every director has his or her own specific meaning for the term.

Warm-up   A preshow performance, usually done by a comedian, announcer, or the show's producer, to make the audience comfortable, to get them laughing, to get them in the mood for the show.

Waveform   Electronic testing instrument that measures and displays picture information.

WGA   See **Writers Guild of America.**

Wild sound   Sound as it is in real life, recorded without scripting or modification.

Window dub   See **Viz-code.**

Wipe   An on-screen picture transition, usually involving some sort of geometric shape or pattern (e.g., a box, a star, a checkerboard).

Wireless (microphone)   A microphone system that uses a radio transmitter and receiver in place of a cable. See also **RF.**

Wrap   The end of a day's work (as in "That's a wrap").

Wraparound   An introduction or closing, frequently used to personalize a movie presentation. Wraparounds are generally prerecorded, and then played back at the appropriate broadcast times to give the appearance that the host is actually watching the movie along with the home audience.

Writers Guild of America   The trade association representing writers who work for the movies, and for radio and television companies under its jurisdiction. Also called the WGA.

Zoom   A lens with multiple focal lengths, permitting easy movement from one focal length (e.g., wide angle) to another (e.g., telephoto), also used as a verb (e.g., "zoom in," or "zoom out").

RECOMMENDED READING

BOOKS

Many recommended reading lists are so long that people don't bother with them. Offered here is a short list of general books, with the hope that the reader will read all of them.

"Must" Reading

Television Production Handbook
Fourth Edition (1984)
By Herbert Zettl
Wadsworth Publishing Company, Belmont, California

This is the best all-around production textbook, and it is available almost exclusively in college bookstores. The book provides detailed information about all aspects of television production, with an emphasis on the technical. This is the book to read if you need to know more about cameras, lighting, audio, videotape, editing, etc. (each has its own chapter). The emphasis here is on "doing it"—so considerable time is spent, for example, on how to operate a ¾-inch VCR in the field. There is a chapter on producing and several on various aspects of directing.

This book is the standard for college courses in television production, and with good reason. With its many photos and illustrations (there are hundreds), it is the finest guide to television production available. Many professionals have learned from it, and still refer to it from time to time.

The Cool Fire: How to Make It in Television
By Bob Shanks
In hardback: W. W. Norton, New York

In softback: Vintage Books, New York
Latest Vintage edition (1983) includes a chapter on cable TV

Bob Shanks is a seasoned professional who has been both a producer and a network executive for many years. He teaches by example, offering personal insights about the nature of the television business in general, and about the specifics of producing and packaging a television production (usually for the networks or for syndication). Chapters are arranged to cover the networks, the syndication business, sponsors, agents, presentations, and ratings, with only two chapters written specifically about production (all of the chapters do include valuable tips about production, however).

Reading this book is very much like interning for a busy producer and packager. Shanks's commentary is frequently revealing of his own personal style and of the style of the business: "Instead of [dining in high style at] "21," there is more often a cold B.L.T. at the desk or in the cutting room or in the rehearsal hall, gobbled down in spasms and flushed away by cardboard coffee. . . ."

Read Shanks's book for the "real world perspective"; it reads particularly well after the "you can do it" philosophy of Zettl (above).

Independent Filmmaking
By Lenny Lipton
Straight Arrow Books, San Francisco
Distributed by Simon & Schuster, New York

Television people who work exclusively with videotape are frequently put off by the mere mention of the word "film." Lipton's book makes film comprehensible. Certainly, video is easier for the newcomer, and it is the medium of the eighties, but film offers some distinct advantages.

Lipton concentrates on 16-mm filmmaking, offering plenty of useful, well-organized information about the practical and the technical sides, and some equally useful opinions. There is a vast amount of information in this book, and it is best read slowly, with plenty of time to absorb in between sittings, and to reread if necessary. It is the best book about filmmaking available, both as a broad introduction and as a reference work.

That's the list. Just three books, including one (Shanks) that is definitely fun to read if you like reading about television. Take the time to read them; they are a good investment.

"Should" Reading

Now, if you want to read more about some specific aspects of television, here are three good suggestions:

Photography
By Barbara Upton and John Upton
Educational Associates, a Division of Little, Brown and Co., Boston

This distillation of The Life Library of Photography is available only in college bookstores. Individual chapters provide remarkably clear, well-illustrated essentials in the use of lenses, cameras, light, color, and film, and the fundamentals of exposure, lighting, color and other topics of certain interest to directors. For more information on any of these areas, visit any well-stocked public library and ask to see the individual volumes from The Life Library of Photography like *Light and Film*, *Photojournalism*, *The Camera*, and *Photographing Children* (the last one is a real treat).

Watching TV
By Harry Castleman and Walter J. Podrazik
McGraw-Hill, New York

Available in most bookstores, this is the best history of network television programming on the market. It follows the action season by season, which permits the authors to show trends and to analyze them with the real world as the background. The story begins with the 1944–45 season (not much of

a season—there were thirteen shows on the four networks), and moves on to the present day, with running commentary, photographs, and lists of milestones for each season.

This Business of Music
By Sidney Shemel and M. William Krasilovsky
Edited by Paul Ackerman
Revised and Enlarged Fourth Edition (1979)
Billboard Books, New York

The subtitle says it all: *A Practical Guide to the Music Industry for Publishers, Writers, Record Companies, Producers, Artists, Agents*. This book is written primarily for people who work in the music business, but it is extremely useful to television producers who work with live music, who are concerned with rights to recorded music, and to producers who work in music video (a fifth edition should cover this field, the fourth does not). Chapters discuss copyright, artist contracts, record producers, labor agreements, licensing, payola, songwriter contracts, performing rights associations, arrangements, public domain music, and so on, with plenty of boilerplate agreements (that can save a lot of legal fees if used carefully). The book can be found in most bookstores and in public libraries. It is also available directly from *Billboard* (see "Magazines" below).

OTHER BOOKS

There are books about lighting, scenic design, makeup, scriptwriting, and performance available in many bookstores and libraries that can be helpful in specific situations. Titles in related areas (motion pictures, theater) can be useful as well.

MAGAZINES

In order to keep up with the latest industry news and trends, many producers and directors read the "trades." There are a few dozen daily, weekly, and monthly magazines and newsletters available to the television and video industries. Each has its audience.

If you are working in entertainment, whether in network, pay TV, syndication, or at a local station, these will be most useful:

Variety
154 West 46th Street
New York, NY 10036

The legendary show business newspaper is a weekly, available either by subscription or at newsstands in larger cities. Motion pictures are still the focus, but there is plenty of space devoted to television, home video, and cable television (as well as theater and live performance). *Variety* provides the best overview of the entertainment business, and it still costs only about a buck a copy.

Daily Variety competes directly with *The Hollywood Reporter*, and tends to be popular almost exclusively in Los Angeles, where it is published.

The Hollywood Reporter
P.O. Box 1431
Hollywood, CA 90078

The daily trade for the Los Angeles motion picture and television industry concentrates on happenings at the studios, at the networks, and the stars. The stories tend to be brief and newsy, sometimes gossipy. Occasional features on home video and new technologies are helpful. The weekly chart of television productions currently under way for network, pay TV, and syndication (listed by company, including addresses, phone numbers, and names of key production personnel) is extremely useful, especially for job-hunters.

Individual copies cost about a dollar, but subscriptions to this daily run several hundred dollars a year.

For those working in local television, there is but one leader:

Broadcasting
1735 DeSalles Street, NW
Washington, D.C. 20036

The weekly report on the state of the television and radio business, with news and features about the networks, sponsors, advertising agencies, national political issues of concern to broadcasters, and so forth. The magazine is written primarily for management, but everybody at the local stations reads it every week. Available by subscription or at newsstands in larger cities.

For those working in cable television, there is:

Cablevision
P.O. Box 5208-TA
Denver, CO 80217

A weekly magazine, also covering the entire cable business. *Cablevision* tends to be more "feature oriented" than *Multichannel News,* and many of these features are about the pay television networks, about programming and programmers, and, occasionally, about made-for-cable productions. There's a monthly bonus section devoted to "business," another to "marketing," and another to "programming," which may prove helpful to imaginative producers and entrepreneurs. *Cablevision* costs a little more than *Multichannel;* for some people, it is worth it.

For news and information about the use of television for education and training, there are several different magazines, with new ones appearing all the time. One of the better ones is *E-ITV,* which is available from Tepfer Publishing, 51 Sugar Hollow Road, Danbury, CT 06810. A visit to an educational or training facility with specific needs will usually unearth a few more titles.

There are several magazines available for those working in corporate television. Once again, the suitability of the magazine will depend upon the orientation of the facility and of the organization. Try *Video Manager* (Knowledge Industry Publications, 701 Westchester Avenue, White Plains, NY 10604), or *Video Systems* (Intertec Publishing Corp., Box 12901, Overland Park, KS 66212).

There are plenty more. Many producers, directors, and executives consider these weekly magazines to be essential reading as well:

Electronic Media
740 Rush Street
Chicago, IL 60611

A complete weekly roundup of developments in network and local television, pay TV, cable, home video, with some information about consumer electronics as well. This publication is relatively new, but it seems to be gaining popularity. If you can afford to subscribe to only one weekly, and need the big stories from *Variety, Broadcasting,* and *Cablevision,* this is probably the best choice.

Billboard
1515 Broadway
New York, NY 10036

Once exclusively a music trade, *Billboard* has become an important home video trade as well. This trend began with its charting of video games and computer software, and continues with its weekly sales and rental charts for videocassettes. A weekly music video rotation chart is indispensable for music video producers. Columns and news stories accompany each of the charts.

A subscription will cost several hundred dollars for the year; individual issues are available at many newsstands. Check your public library (or college library) as well.

Advertising Age
740 Rush Street
Chicago, IL 60611

The comprehensive weekly journal of the advertising business, offering roughly half of its space to activities in television (and radio). *Adweek,* a competitor, is also a good choice (820 Second Avenue, New York, NY 10017). Both are weekly trades and cost a few hundred dollars a year for subscriptions.

If you're still hungry for information about media, there are several more recommendations. Many directors enjoy reading *Videography* (P.O. Box 658, Holmes, PA 19043), which covers the entire production trade, updates developments, and offers regularly scheduled roundups of equipment, facilities, and other useful information for broadcast and nonbroadcast television. *Millimeter* (826 Broadway, New York, NY 10003) covers both the television and the motion picture production businesses.

Producers, particularly those who buy the rights to books for motion picture adaptation, read *Publishers Weekly* (R. R. Bowker Company, P.O. Box 1428, Riverton, NJ 08077) for reviews of the latest releases from the book publishers. *PW* is becoming quite popular among producers of home video programs as well.

The television trades tell only a part of the story. It is essential that a producer (and, to a lesser extent, a director) keep aware of events in the real world. Reading the *New York Times,* the *Wall Street Journal,* and *Business Week* regularly is essential for some, while *People* magazine, *USA Today,* or a local "city magazine" (e.g., *Washingtonian*) might be more useful for others. Television is a business of ideas, and many of these ideas are sparked by reading magazines.

Index